Mushroom Day

On this mushroom-filled day, ecologist Alison Pouliot tours the world to introduce readers to a fascinating variety of fungi. Each chapter of *Mushroom Day* introduces a single fungus during a single hour, highlighting twenty-four different species. In the dark of the night, the green glow of the ghost fungus guides us into the forest to learn about the mysteries of bioluminescence. At dawn, we awaken to find a fairy ring of mushrooms that has appeared overnight like something out of folklore. Later in the evening, we spy a fungus known as the witches cauldron and wonder what it might tell us about the future. By the end of our mushroom day, we'll have glimpsed the diversity of this unique kingdom, met fungus friends that feed and fascinate, and learned how humans can encourage their flourishing.

For each hour, celebrated artist Stuart Patience has depicted these scenes with evocative pen and ink illustrations. Working together to narrate and illustrate these unique moments in time, Pouliot and Patience have created an engaging read that is a perfect way to spend an hour or two—and a true gift for foragers, mycophiles, and anyone who wants to stop and appreciate fungi.

Praise for **MUSHROOM DAY**

"Wow! What a cool way to learn about fungi. Each of the twenty-four mushrooms described here is a gateway to an aspect of mycology: lion's mane and cultivation, the ghost fungus and bioluminescence, the veiled polypore and symbiosis. In a style both easygoing and thoughtful, Pouliot delivers enough science to be accurate but not so much that a newbie will stumble—no easy feat. While this book will delight folks at all stages in their fungal journey, it will be especially inviting to those just starting out. And Stuart Patience's graphic illustrations? Simply gorgeous. *Mushroom Day* is in every way a great addition to any mycophile's library."

EUGENIA BONE, author of *Have a Good Trip: Exploring the Magic Mushroom Experience*

"A feast of fungal wonders and superb drawings. *Mushroom Day* is a fabulous illustration of symbiosis: between fungi and animals, fungi and plants, science and art—and more— with creativity at its core. A delight!"

GIULIANA FURCI, founding director, the Fungi Foundation

"From the luminous to the ghoulish, the fragrant to the stinky, the tiny to the giant, Pouliot showcases twenty-four very different fungi—often taking the reader to far-flung places across the globe to find habitats where these special fungi live. And no matter the time of day or night, there's a fungus of relevance. Whether fungi are entirely novel to you or you are already a fungus fan, enjoy the adventure of discovering this often-overlooked world, as revealed through this intriguing journey through the second-largest kingdom of life."

PETER BUCHANAN, Manaaki Whenua – Landcare Research (Aotearoa—New Zealand)

"*Mushroom Day* is popular science at its very best— knowledge spiced with passion for the fungal kingdom. The combination of evocative and diverse stories of fungal wonders is captivating."

ANDERS DAHLBERG, Swedish University of Agricultural Sciences

MUSH ROOM DAY

A STORY OF 24 HOURS AND 24 FUNGAL LIVES

WRITTEN BY	ILLUSTRATED BY
Alison Pouliot	Stuart Patience

The University of Chicago Press
Chicago and London

The University of Chicago Press, Chicago 60637
The University of Chicago Press, Ltd. London
© 2025 by Alison Pouliot
Illustrations © 2025 by Stuart Patience
Published 2025
Printed in Türkiye

34 33 32 31 30 29 28 27 26 25   1 2 3 4 5

ISBN-13: 978-0-226-83844-1 (cloth)
ISBN-13: 978-0-226-83845-8 (ebook)
DOI: https://doi.org/10.7208/chicago/9780226838458.001.0001

Library of Congress Cataloging-in-Publication Data

Names: Pouliot, Alison, author. | Patience, Stuart, illustrator.
Title: Mushroom day : a story of 24 hours and 24 fungal lives / written by
    Alison Pouliot ; illustrated by Stuart Patience.
Description: Chicago : The University of Chicago Press, 2025. | Includes
    bibliographical references and index.
Identifiers: LCCN 2024058223 (print) | LCCN 2024058224 (ebook)
    | ISBN 9780226838441 (cloth) | ISBN 9780226838458 (ebook)
Subjects: LCSH: Mushrooms. | Fungi.
Classification: LCC QK617 .P669 2025 (print) | LCC QK617 (ebook)
    | DDC 579.6—dc23/eng/20250117
LC record available at https://lccn.loc.gov/2024058223
LC ebook record available at https://lccn.loc.gov/2024058224

♾ This paper meets the requirements of ANSI/NISO Z39.48-1992
(Permanence of Paper).

Contents

Preface

For a long time, people thought mushrooms were plants, but they're not. They're not animals either. They're fungi—or more precisely, they're the sporing bodies of some fungi. And fungi differ from plants and animals in crucial ways. Unlike animals, they can't run or fly away when things get rough; they must respond to their environment to survive. Unlike plants, they don't have chlorophyll and so can't use light to make their own food; instead they use enzymes to digest proteins, carbohydrates, and fats. They're almost like animals in that way, but fungi don't have stomachs: they live within their food, such as twigs and sticks, leaves and logs, live animals and dung.

Fungi are similar to animals in other ways as well. Their evolutionary branches run parallel and go back to a common ancestor. Think of it this way: Fungi and animals are different life-forms that essentially do the same thing. They're just designed differently, a little like iPhones and Androids. Fungi also share specific traits with some animals. Insects and crustaceans have a structural component in their exoskeletons called chitin, the same compound that fungi use to build cell walls (plants, on the other hand, use cellulose). And some fungi even do something that we humans do: produce vitamin D when exposed to sunlight.

Although fungus reproductive structures, like mushrooms and other sporing bodies, have simple forms, they're remarkably diverse. The umbrella-shaped mushroom is perhaps the best known among them, but we'll discover others during our mushroom day, all of which serve the same sole purpose of dispersing spores. While many of us can name the parts of a flower, fungus structures are often less familiar. To get started, a mushroom has a cap called a pileus and a stalk called a stipe. On the underside of the pileus (known as the fertile surface or hymenium), we'll often find lamellae (gills) or pores that release spores.

In this book, we'll meet twenty-four species over twenty-four hours, each with its own unique character and stories. The sun might rule a bird's or frog's day, but can a fungus tell time?

Fungi can certainly sense light—they have up to eleven types of photoreceptors that enable them to detect frequencies from near-ultraviolet to near-infrared light. Many fungi need at least some light to grow and reproduce. The sun helps a mushroom know which way is up and which way is down, and that's vital for getting its spores out into the world. But be warned: Fungus reproduction is not for the fainthearted, nor is it simple. Ideas about the male-female dichotomy start to wobble when we consider that some fungal innovators have more than twenty-three thousand mating types. And while most fungi reproduce by mating (through sexual reproduction), many can also reproduce without mating (through asexual reproduction). Although it may be the showy sporing bodies that draw us to fungi, most species don't bother making spore vessels at all and instead produce asexual spores directly from their mycelia in the secrecy of the subterrain. What's a mycelium? It's the fungus feeding body, made up of a matrix of threadlike fungal cells called hyphae, that explores the soil in search of food. Mushrooms and other sporing bodies are just the "organs of the organism," or the fungal equivalent of a plant's flowers or an animal's gonads.

Being able to reproduce both ways is a smart strategy for adapting to environmental conditions. When conditions are favorable, asexual reproduction allows a fungus to spread quickly. But when the environment changes or becomes unstable, genetic variation is key to survival, and sexual reproduction provides the advantage of new arrangements of genes.

Mushrooms respond to the full spectrum of light, but certain wavelengths trigger different responses. Blue light, for example, is necessary for the fungus mycelium to develop, while red light stimulates mushroom growth. Some fungi, like humans, need darkness to rest, and others need light to defend themselves—for example, by producing melanin as a sunscreen. Light also influences the production of compounds that affect nutritional value and flavor. Farmers, therefore, will not only mimic natural cycles of light but also experiment to maximize yields and achieve desirable culinary traits. By exposing mushrooms to different light regimes, they can help a fungus grow in the right direction and reach optimal size, shape, and quality.

To return to the important question: Do fungi follow a daily clock, a circadian rhythm, like we do? Circadian clocks orchestrate an organism's daily cycles and generate self-sustaining rhythms. Because the earth rotates and exposes us to cycles of day and night, our circadian clocks allow us to anticipate and adapt to these changes. They help us survive. Our understanding of fungal responses to light comes mostly from microfungi, like molds, rather than macrofungi, like the mushrooms on your pizza. For over fifty years, researchers have used an unassuming bread mold called *Neurospora crassa* to observe rhythms of growth and asexual spore production and to investigate how biological clocks are regulated in fungi. They have discovered that many of the same molecules that underlie this fungus's circadian clock are also present in animals.

Although many fungi can detect and respond to light, I'm not sure that I would say they have "fungal vision." That might be taking it a little too far, and besides, I hope to continue enjoying my time in the forest without feeling like the mushrooms are watching me! More importantly, by spending that time researching and protecting their habitats, we can discover the great diversity of fungi working away in the subterrain, helping to hold the earth's ecosystems together.

Relative to animals and plants, fungi still conceal many mysteries from us. Limited knowledge of fungal biodiversity is especially worrying in a rapidly changing world. The recent ground-swell of interest in fungi, however, suggests that mushrooms are having their moment. Our understanding of this remarkable kingdom of organisms is growing, helping mycologists—or fungus experts—better understand their needs and how to support their flourishing. And to that effect, we can certainly punctuate our day with fungi. The timing of our mushroom meetings in this book is as much about the daily routines of the fungus foragers, both creature and human, as it is about the fungi themselves.

We begin at midnight with the midnight disco, which is not a late-night dance club but a rare fungus. In the wee hours, the curious Australian ghost fungus lights our path with its eerie green glow. At dawn, we awaken to find a fairy ring of mushrooms that has sprung up on the lawn overnight. At 7 AM, it's time to get to the forest in the Piedmont region of Italy, before other porcino hunters forage the fattest and finest fungi. The lawyer's wig is another tasty mushroom, but it has an unappetizing habit of melting into an inky mess. In the heat of the late afternoon, the stench of the stinkhorn could well repel you from the forest, after the enticing aroma of the aniseed funnel had lured you in. As the evening draws near and we gather around a campfire to roast our foraged chestnuts, you might discover a tiny fungus—the hairy nuts disco—hiding out in the chestnut's inner husk.

At the end of the day, we'll have glimpsed the diversity of the kingdom Fungi, discovered some of its members' quirks and traits, and realized that we can't exist without them. I hope this global foray piques your interest in the mycological miracles that surround us all day, every day.

ALISON POULIOT
Mount Franklin, Australia, and Bienne, Switzerland

Artist's Note

The presence of the extraordinary among the mundane—a phenomenon that fungi encapsulate so powerfully—fascinates me. We are surrounded by these incredible organisms, hidden just beneath the surface or sometimes right before our eyes, but we're often too preoccupied with everyday life to notice them.

Mushroom Day awakened me to how remarkable and diverse fungus species are. Some appear deceptively small and unassuming, as just the tips of vast networks of rootlike mycelia that can grow thousands of acres across our forests. Others look so otherworldly that you'd expect to find them on an alien planet.

At the very start of this collaboration, Alison and I discussed how we wanted the drawings to not only capture fungus anatomy but also reflect her evocative stories. I like to inject an element of folklore into my illustrations—sometimes by hiding clues in the background that might shed some light on the subject without being too literal. I hope that these clues pique your interest and that you see them as invitations into the incredible world of fungi.

We also wanted to depict the light of a passing day, beginning with the pitch blackness of midnight and brightening gradually before time dusked it again. The limitations of black and white pushed me to be more inventive with my use of tone, pattern, and perspective.

Working with Alison on this book has been such a joy. She is an encyclopedia of knowledge, and her infectious enthusiasm inspired me to illuminate these magical fungus worlds. Next time I go out strolling in the forest, I will be sure to stop and look more closely at my surroundings, hoping to discover one of these twenty-four magnificent species.

STUART PATIENCE
London, England

A Note on Names

Fungi, like many organisms, often have more than one common name. The vignettes in this book use the local common name alongside the current scientific name, which is the same worldwide. Not all fungi have common names, in which case only the scientific name is provided.

Mushroom Day

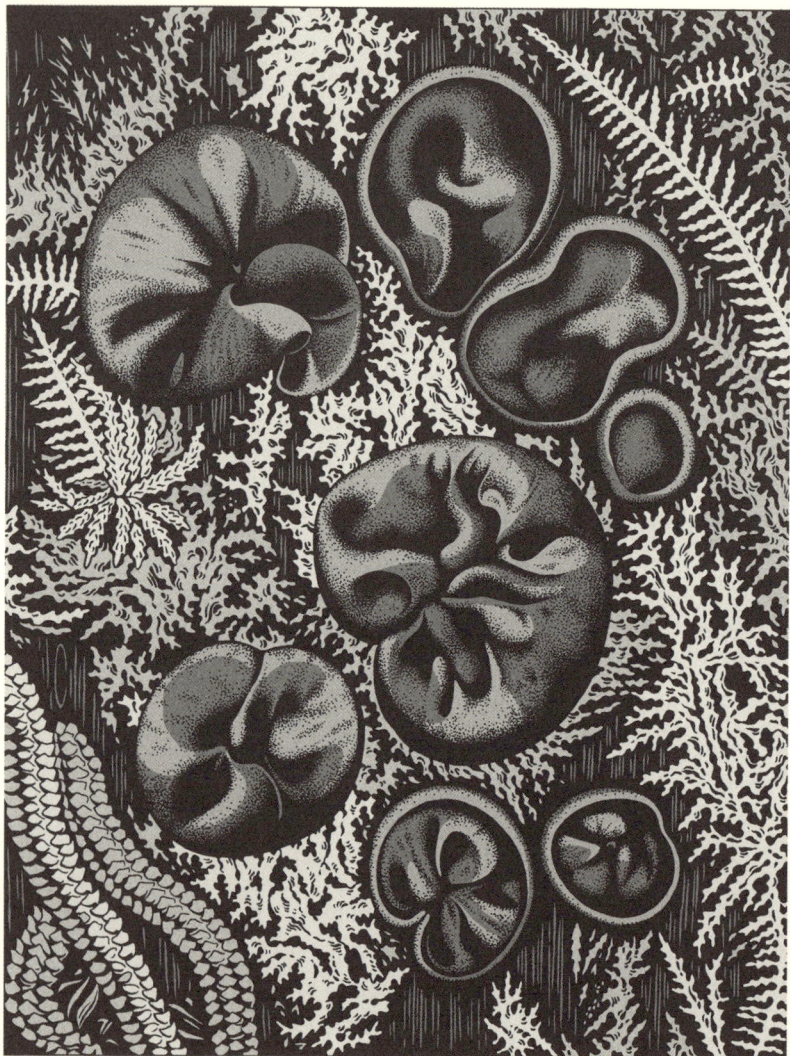

Midnight Disco

Pachyella violaceonigra

(EUROPE)

The midnight disco may well be a late-night dance club, but it's also a rare fungus. Given its name, it serves as a perfect species both to start our day and to consider the timing of life more generally.

Like most fungi, the midnight disco is a saprotroph, or recycler, obtaining its food by secreting digestive enzymes directly into the organic matter in which it lives. But the midnight disco is also unusual: Unlike most fungi, which produce mushrooms and other sporing bodies in the fall, it pops up in the spring and sometimes in the summer. Ecologists study a fungus's phenology, or the seasonal timing of events in its life, to better understand how it operates within its ecosystem. A fungus's access to resources, such as food and water, depends on the timing of phenological events. The midnight disco, as well as some other saprotrophic fungus species, may have adapted to spring fruiting to minimize competition for resources with other fungi.

Although mycologists and mushroom hunters are skilled at anticipating when a fungus will produce its sporing bodies, predicting when they'll emerge is becoming more difficult as the climate and seasons change. For example, in alpine regions, snow is melting earlier, and spring is getting a head start. Some organisms adapt by bringing forward their spring activities, such as reproduction, while others are slow to change with their environment. These different responses mean that some species miss important meetings: Plants may not flower in sync with pollinators, or predators may go hungry if their prey don't appear when expected. As we'll see throughout the day, fungi also rely on the phenology of plants and animals, especially those that share resources or help distribute spores. A change in the timing of one species' activities can trigger a cascade of consequences for others.

That's one reason the midnight disco is found on Red Lists, inventories of species' conservation status. Red Lists were first developed by the International Union for Conservation of Nature, which has been tracking the risk of species' extinction for sixty years. Red Lists largely neglected fungi until 2015, when a group of dedicated mycologists developed the Global Fungal Red List. This list gives us an international overview of different fungus species' risk of extinction, the various threats to their existence, and their distribution and ecology. Scientists assign each fungus on the list a category, such as "vulnerable" or "critically endangered," that indicates the degree of threat and allows them to prioritize conservation efforts. Although fewer than a thousand fungus species have been assessed for the list, representing only a tiny fraction of the estimated five million fungus species, it's a start. The list is a valuable tool for conservation and helps bring fungi, particularly rare and endangered ones, into the spotlight.

Speaking of spotlights, why is this fungus called the midnight disco? The specific epithet *violaceonigra* refers to the violet-black color of the inside of the sporing body's "cup," although the color can vary from purplish black to purplish brown to reddish black. The first part of the fungus's common name thus alludes to its generally "midnight" color rather than the time of day when it appears, while the second part is an amusing shortening of its former scientific class, Discomycetes.

The midnight disco produces sporing bodies that can grow as big as an eggcup. Special saclike structures called asci line the interior of the cup. These are where the spores form. As the spores mature, pressure builds within the asci until their tips burst, launching the spores into air currents above. This propulsion helps get them further afield than if they relied on wind alone. Some mycologists think that the cup shape also captures raindrops that may help fling out the spores. In the same way that plants produce an enormous variety of flowers to maximize their reproductive success, fungi are ingenious innovators of sporing-body design, employing every trick in the book to spread their spores far and wide.

More species find themselves in trouble as the world changes and their habitats disappear or decline in quality. Plants and animals have been the focus of conservation programs worldwide, but fungi are gaining more attention. Again, this comes down to timing. Seasonal, ephemeral, and often unpredictable, fungi are difficult to survey and monitor, especially inconspicuous ones like the midnight disco. The rise of molecular mycology, new conservation tools, and citizen contributions to online biodiversity repositories, however, is increasing our understanding of where species grow and how they live. These collaborative efforts will hopefully ensure that it's not the last dance for the midnight disco.

1 AM

Ghost Fungus
Omphalotus nidiformis

(AUSTRALIA)

On a moonless night, if you let your eyes adjust to the complete darkness, the eerie pale-green glow of the ghost fungus will draw you in like a moth to a flame. As they come into view, you'll notice that these large and impressive mushrooms are everywhere. Look down, and you'll see them surrounding the bases of living trees and dead stumps. Look up, and you'll spot them protruding from trunks and limbs. And although their glow is subtle, these ghosts can shine bright enough to allow you to read this book.

Endemic to Australia, the ghost fungus grows in a range of habitats from eucalypt forests to

pine plantations, rainforests, and coastal scrub. Although it appears on trees, it's a saprotroph (recycler), not a parasite (organism that exploits a host). Its identification can be tricky, as it appears in an astonishing array of colors: Young ones can be a matte bluish black; others, reddish brown; and the occasional one, mauve or even greenish. But mostly, these fungi are white, often splotched with orange or other colors. As they age, the funnel-shaped mushrooms buckle and contort, and scallop along their edges. But it's in the darkness that this glowing fungus is most entrancing.

Scientists describe the ghost fungus as bioluminescent—from *bios*, meaning living, and *lumen*, meaning light. Like all biolumi-nescent fungi, it emits a low-wavelength light, which appears faintly green. Human eyes can't detect this phenomenon in daylight, but the ghost fungus glows continuously nonetheless. Bioluminescence occurs across kingdoms of organisms, and we've known about it for centuries: It features in age-old Buddhist texts and ancient Chinese poetry. The Greeks also noted marine phosphorescence over 2,500 years ago, and we have a pretty good idea as to why other organisms glow. But despite mycologists' efforts to reveal its secrets, the ghost fungus keeps us guessing.

Long before the scientific investigation of bioluminescent fungi, the ghost fungus likely appeared in the stories of Australia's First Nations peoples. First Nations peoples—like all cultures—respond to fungi in different ways: Some use them to their benefit, while others avoid them. Several species are eaten, others are used medicinally, and still others are used as dyes, as tinder to start a fire, or to repel flies. Early European settlers in Australia reported that some First Nations peoples were fearful of what is believed to have been the ghost fungus. Two Aboriginal names for it, *Chinga* and *Mettagong*, mean "spirit," and the people of at least one nation in Western Australia associated its luminosity with campfires of departed evil spirits. Yet other First Nations peoples were unafraid of and even amused themselves with these glowing marvels.

So, what makes the ghost fungus glow? It contains a substance called luciferin, which was probably named after Lucifer, the light bearer in classical Western mythology. In the presence of oxygen, luciferin undergoes a chemical reaction that causes the mushroom to release energy as a pale, greenish light. The soft-winged beetles known as fireflies produce light using a similar chemical reaction; each species has its own special flashing pattern for recognizing kin and finding mates. But unlike those insects, ghost fungi don't glow to attract mates.

Why glow at all? Some bioluminescent fungi attract spore-dispersing insects, helping to ensure future fungal generations. Tempted by that hypothesis, mycologists tested the ghost fungus's attractiveness to mosquitoes and found that the species does not draw more insects than nonbioluminescent fungi. The magical light may simply function to remove by-products of daily life. Animal mechanisms do this too. We sweat to remove excess heat and burp to release surplus swallowed air. If imagining the ghost fungus glow as the equivalent of a human belch is unsatisfying, being the creative species that we are, we'll no doubt continue to come up with stories to account for this remarkable phenomenon.

A few years back, I found an impressive flush of ghost fungi in a pine plantation on the southeast coast of Australia. Hundreds of ghosts in huge clusters adorned the bases of the trees. I collected a few and took them to a fungal ecology workshop that I was running. The workshop was held in a church, and near the entry was a small, windowless room. Pitch-black, it was the perfect place for the participants to witness the glow. I placed the bioluminescent beauties on a stand, and one by one, the participants filed in. Each waited a moment for their eyes to adjust to the darkness, then let out squeals of delight. Without a doubt, the ghost fungi were the highlight of the day. The following morning, I was a good three hundred miles further up the coast when I realized that I'd forgotten the ghosts. Left to haunt the church, they surely started to putrefy. Oops! But at least from within the darkness, each observer had seen the light.

2 AM

Honey Fungus
Armillaria mellea

To appreciate the honey fungus, we need to dig.
The species, which is both parasitic and saprotrophic, has a glowing mycelium that wends and
winds its way through wood and soil. How would
a fungus benefit from glowing underground?
A young friend of mine suggested that the
light might make it easier for worms to tunnel
through the soil. It's a charming suggestion, but
worms can navigate on their own. Both honey
fungi and worms have adapted well to life in the
darkness.

We know from the ghost fungus we just met
that bioluminescence is a complex phenomenon,

evolving many times in the history of life and for different reasons. Researchers have identified an astonishing eight hundred genera of organisms worldwide that bioluminesce. While various beetles, flies, centipedes, millipedes, and snails have this remarkable ability, bioluminescent organisms are far more common in marine environments. In the darkest depths of the ocean, the female anglerfish employs a luminous lure of glowing tentacles to ensnare her unsuspecting prey. Other creatures use tricks of light to ward off potential predators. Some marine algae and bacteria cause "glowing seas" that have colored sailors' stories across the centuries.

We know that eighty different species of fungi can bioluminesce. Many of these are tropical species, but some grow in temperate regions. Among the various bioluminescent fungi, honey fungi are odd in that it's their mycelia, not their mushrooms, that glow. Although these gossamer-thin cylindrical cells are individually microscopic, we can see them en masse. For a long time, scientists thought all honey fungi were a single species, *Armillaria mellea*. But because they vary a lot in appearance, mycologists dug deeper and discovered that there are at least eleven distinct species of honey fungi, eight of which glow.

The honey fungus lives in wood, and its glowing mycelium can create the impression that the wood itself is luminous. In the early nineteenth century, the wooden structural beams of a coal mine in Germany's Ruhr valley were streaked with light bright enough for the miners to work without lamps. A century later, British mycologist John Ramsbottom described how World War I soldiers attached glowing wood to their helmets and rifles. This biological light source presumably helped them navigate the trenches while not shining brightly enough to attract the enemy. Despite the creative ways humans have put the honey fungus's brilliant mycelium to use, we still don't understand why it bioluminesces. As for the ghost fungus, bioluminescence is probably a detoxifying by-product of metabolism and doesn't provide any advantage for reproduction.

Along with the ability to bioluminesce, the honey fungus has another remarkable trait—it can grow extraordinarily old and large. The mycelia of some honey fungi are several thousand years old. Although the familiar phrase "everything is bigger in Texas" often rings true, in 1998 no fungus—indeed, no known living organism—was larger than the giant honey fungus (*A. ostoyae*) discovered in eastern Oregon. Researchers later found other massive honey fungi in Washington and Colorado. In 2003, however, Oregon won back the prize with another specimen recorded at a whopping 2,384 acres (that's almost the size of London's Heathrow Airport) and estimated to be at least 2,400 years old. This monstrous fungus—which is probably not a continuous single mycelium of living cells, all in communication with each other, but rather the extent of fragmented parts of a colony with the same genes (clones)—raises interesting questions about what constitutes an individual organism. And it displays another remarkable strategy for survival, ensuring there's a spare set of genes nearby.

The honey fungus has a special feature that allows it to grow so large and gain so much territory—rhizomorphs. These cords of parallel hyphae, which resemble shoelaces, give the honey fungus its alternative names: bootlace or shoestring fungus. Rhizomorphs can transport water and nutrients over long distances and infect and decay more trees. They also contain melanin, which helps protect them against ultraviolet radiation and extreme temperatures and keeps them from drying out as they move from tree to tree. If you peel back the bark of a fallen tree, you might see the black weblike weaving of rhizomorphs.

Mycologists suggest that the honey fungus's ability to reach such an impressive age and size could also relate to the changing conditions of forests. Forests today are more disturbed and consequently have a different composition of trees, with fewer older ones. Young trees are often more vulnerable to disease, so these changes could favor the spread of honey fungi and restrict the other fungi that once kept them in check by competing for resources.

I like to contemplate how a microscopic spore can grow into such a large organism while remaining underground and out of sight. The honey fungus reminds us that much more is happening in the subterrain than we realize—and that fungus discoveries can brighten any hour.

Devil's Bolete

Rubroboletus satanas

(EUROPE)

Waking up in a cold sweat, mushroom forager Simone Sylvestre is roused from slumber by a nightmare about the devil's bolete. This imposing mushroom is conspicuous, even daunting, and it appeared in her dream in all its sinister splendor.

Boletes are mushrooms with an underbelly of pores or tiny holes. Many boletes are edible, and foragers seek species like the famed porcino (*Boletus edulis*), which we'll meet later. The devil's, or Satan's, bolete is an exception. If its lurid bloodred underbelly and knack for turning blue when handled are not off-putting enough,

then edge up close to an older one; its putrid stench should have you convinced. This striking mushroom is one to admire, not to consume, or you'll end up violently ill. You might suffer awful abdominal pain, vomiting, and bloody diarrhea, along with dizziness and dehydration. Worst of all, this unpleasant and unforgettable combination of symptoms could torment you for up to six hours.

The culprit responsible for that toxic effect is a compound called bolesatine. Bolesatine is a large protein comprising over five hundred amino acids. In high doses, it has the even more dire effect of inhibiting protein synthesis. That's not good because our bodies manufacture proteins as enzymes that are vital to the structure, function, and regulation of tissues and organs. Tests on luckless rodents showed that bolesatine caused liver damage and thrombosis (blood clotting). Another toxin, muscarine, which is found in fungi like fibrecaps (*Inocybe*) and funnels (*Clitocybe*), is also present in the devil's bolete but in quantities too small to cause a toxic effect. Although severe poisoning from bolesatine could be fatal, symptoms of eating this fungus are usually gastrointestinal, and as far as we know, no one has died from its consumption.

German naturalist Harald Othmar Lenz first described this fungus in 1831 and called it *Boletus satanas*. It received its current name in 2014. Lenz not only named the species but attested to its rumored toxicity by sampling the fungus himself, becoming ill and reputedly suffering from weakness, vomiting, and seizures. His choice of the specific epithet *satanas* might prompt a potential consumer to reconsider. Going beyond the call of duty of a mycologist, Lenz may well have saved lives, or at least reduced unpleasant poisonings.

The devil's bolete grows in mixed woodlands in Europe, especially in the warmer southern regions. Although widely distributed, it's considered rare. Unlike the saprotrophic and parasitic fungi we've met so far, the devil's bolete is mycorrhizal. Mycorrhizal fungi feed by forming mutually beneficial relationships with plants, attaching to their roots and trading nutrients and water for sugar. The devil's bolete forms its relationships with various broad-leaved trees. You're likely to find it among mature beech, oak, chestnut, hornbeam, or lime trees—most often in limestone or chalky soils. I first encountered this gargantuan fungal wonder nestled in leaf litter in the lowland beech woods in southern England. Although it was our first meeting, the fungus's portly posture, garish coloring, and lingering fetor revealed the devil's true nature.

Mushrooms change in appearance as they mature. The cap of the devil's bolete starts out grayish white and develops tinges of olive brown. It is tomentose, or downy, when young, becoming smooth and a little sticky as it ages. It can reach the size of a bicycle helmet, sometimes even larger. On its underside, the rounded pores are yellow to orange at first, red soon after, then purplish red or deep carmine at full maturity.

When bruised, they turn blue. Looking under the cap, you'll see that it's supported by a thick stipe often as fat as it is tall and decorated with a fine, yellowish to crimson hexagonal net known as reticulation.

People sometimes assume that the color red is a warning. When it comes to fungi, however, color bears no relationship to edibility or toxicity. This red-pored fungus is indeed toxic, but there are also pale and nondescript mushrooms that are deadly. Mycologists named these species after evil mythical figures because of their toxicity, not because the devil supposedly wore scarlet or any such legend. Risks to heedless *Homo sapiens* aside, it's astonishing that some fungi can concoct such a toxic cocktail of chemicals. Why do they produce those compounds? We don't know for sure, but it could be a defense mechanism to protect themselves from being eaten by animals before they've dispersed their spores.

At 7 AM, Simone will be roused again—this time by her alarm. As she gets up and dressed, the faint recollection of a nightmare will unsettle her, but she won't recall the details. She'll pause for a moment, until something about the dream prompts her to pop her fungus field guide into her basket. She'll then pull on her boots, whistle for her dog, Lucky, and head for the forest.

Ergot

Claviceps purpurea

(WORLDWIDE)

A baker starts the day early, kneading dough and baking bread before the sun rises. So it was in the Middle Ages, when a mysterious toxic fungus that colonizes cereal grains found its way onto the breakfast table, changing the course of human history and agriculture and leading to devastating epidemics.

Claviceps purpurea is the most notorious of the forty-odd species of ergot fungi. Like the others, this species is a plant parasite. But unlike most *Claviceps*, which have just a few hosts, it infects over four hundred plant species, including sedges, rushes, and grasses, as well as many important cereals, like wheat, oats, and rye. Having a range of hosts allows a fungus to spread far and wide, especially when humans plant a single species alone, as is the case with monocultures of cereal crops.

The airborne ergot spores first colonize a plant host, such as rye, while it's flowering. A sugary, spore-containing slime called honeydew forms, and its foul odor draws dozens of different insects, like flies, beetles, and wasps, which feast on the sticky slurry and then fly off to healthy flowers that too become infected. Rain and wind also contribute to spore dispersal, and cool and damp conditions favor the flourishing of the fungus.

Ergot hijacks the food supply that the rye needs for seed production. About two weeks after the initial infection, the fungus's mycelium hardens and replaces the individual rye kernels with purple "resting structures" called sclerotia. Each sclerotium protrudes from the ripe rye ear like a spur—or *argot* in French, from which the name *ergot* comes.

Sometimes medieval harvesters and millers accidentally mixed ergots with healthy grain. The bakers would then unwittingly add the fungus to their dough. The ergots would survive the baking, and those who consumed the bread made from the infected flour would experience dramatic and frightening symptoms: twitches and spasms, headaches, hallucinations, and paranoia. Ergot poisoning, or ergotism, also hinders blood flow, leading to a loss of peripheral sensation, gangrene, and the death of body tissues. Also known as St. Anthony's fire or the devil's curse, this condition left thousands across Europe and the United States dead or disabled. There was little understanding of medicine, let alone mycology, in the Middle Ages, and such startling symptoms seemed inexplicable. This may have contributed to the accusations of bewitchment that saw thousands of people put to death. It was a long time before doctors associated the fungus with the maladies.

Humans weren't the only species who suffered the effects of ergot. Grazing stock ate ergots that fell to the ground. Those cows and sheep then faced symptoms like reduced fertility and lactation and dry gangrene in their hooves. Ergots have a tough exterior, and those that weren't eaten survived the winter. In the spring, they released their spores, starting the cycle over with the new season's crop.

The symptoms of ergotism result from a toxic cocktail of diverse alkaloid chemicals. The illicit and powerful psychedelic drug LSD (lysergic acid diethylamide) is derived from these same alkaloids. Like the plants that produce alkaloids such as morphine and quinine, ergot provides the raw materials for drugs that can be beneficial and therapeutic at times and harmful at others. As the classic toxicology maxim given to us by the Swiss physician and alchemist Paracelsus attests, "sola dosis facit venenum"—the dose makes the poison.

As early as the 1500s, observant midwives noted how ergots induced premature labor in pregnant sows and used this knowledge to speed up stalled labor in women or initiate abortions. The midwives were, in fact, observing the effects of an alkaloid called ergometrine. The use of ergometrine became more mainstream in the 1800s, but the treatment was risky because of difficulties in determining dosage. This led to an increase in stillbirths. Researchers later purified ergometrine, allowing it to be administered with greater control. Today, doctors use safer drugs to induce labor but may still use ergometrine to prevent uterine bleeding during and after child-birth.

The dual identity of ergot as a source of both dangerous toxins and beneficial pharma-ceuticals, especially for diseases of the central nervous system, persists. Migraine sufferers use ergot alkaloids, and medical researchers are assessing their efficacy in treating degenerative illnesses such as Parkinson's disease.

Although ergot continues to infect cereal crops, scientists and growers are working to improve both crop resistance and the management of contaminated grain. Changes in cultivation methods—from crop rotation and deep plowing to the physical sifting of sclerotia out of grain and the application of fungicides—are well underway. There is no silver bullet for combating ergot, however, and stock animals still occasionally fall ill after eating infected grain. So long as humans produce grain in large quantities, the fungus will continue to influence agriculture and history.

Hou Tou Gu
Hericium erinaceus

(ASIA, EUROPE, NORTH AMERICA)

As for the bakers we just met, it's an early day for mushroom workers in Hailin city in northeast China's Heilongjiang province. They're producing the stunningly beautiful tooth fungus known as hou tou gu.

Hou tou gu is not your average mushroom. For a start, it doesn't have the familiar cap-and-stalk configuration. Instead, this elegant fungus appears as downward-dangling white spines. Imagine a miniature frozen waterfall or clusters of spindly stalactites, and you're getting close to an impression of this unusual fungus.

Hou tou gu's generic name *Hericium* means

"hedgehog" in Latin. The species' many common names reflect its likeness to those spiny creatures, as well as its associations with other animals. In English it's called the hedgehog or bearded hedgehog hydnum, lion's mane, bear's head mushroom, monkey head mushroom, or pom pom. The French also call it the hedgehog (*hydne hérisson*), as do the Swedes (*igelkottstaggsvamp*) and the Germans with a slight variation (*Igelstachelbart* or "hedgehog goatee"), while the Dutch call it the wig mushroom (*pruikzwam*). The Japanese attribute to it a more spiritual significance and refer to it as *yamabushitake*, which means "mountain priest mushroom."

Mushrooms have been a fall staple in China for centuries. Hou tou gu has a long-standing reputation as both traditional food and folk medicine, not only in China but also in Japan, India, and South Korea. Many Asian people use it as a tonic to support overall health and longevity and to promote their Chi, or "life force." The Chinese began cultivating mushrooms over a thousand years ago (starting with shiitake, *Lentinus edodes*), although hou tou gu cultivation began only around 1988. Today, it's one of the five most popularly cultivated mushrooms worldwide. Japan produces the largest quantity, followed by China, South Korea, Taiwan, the United States, and Canada.

New techniques and technology have revolutionized commercial mushroom production since its humble beginnings. It's now a multibillion-dollar industry. A variety of mushrooms is available, and the demand for a still greater range of species is growing—especially from Southeast Asian and Western markets, where consumers regard them as both a meat alternative and a health benefit. Hou tou gu, in particular, is attracting new buyers because of its alleged cognitive benefits.

Hou tou gu purportedly combats a wide range of ailments. Some claim that it is effective against cancer, stomach ulcers, diabetes, and heart disease. Others say that the fungus supports the immune system; improves liver function, nerve development, and brain function; and reduces inflammation. Still others prescribe it to fight viruses and bacterial infections or to lower cholesterol and reduce blood clots. It's used to combat depression, dementia, and other degenerative brain diseases; reduce anxiety; and ease symptoms of menopause. The list of treatments is impressive, and while there are some promising indicators of the fungus's effectiveness, there is little direct evidence to support these assertions.

The many paradigms of addressing human health don't always align. Conventional Western medicine and traditional Eastern medicine have different approaches to testing, diagnosing, treating, and preventing disease. The challenge for consumers is to distinguish evidence-based claims from pseudoscience and marketing. Meanwhile, commercial products from fresh and dried mushrooms continue to proliferate: liquid extracts, powders, tinctures, and even "brain fuel chai latte" mix, which purports to be a "superblend of hou tou gu to enhance cognitive alignment and calm the mind." Hou tou gu hyperbole abounds, but a walk in the forest searching for the fungus, regardless of whether you find it, is likely to reduce your stress and boost your fitness and happiness.

In the wild, hou tou gu grows in temperate forests throughout the northern hemisphere. We find it on living trees as well as large logs, stumps, and stags. Although it is sometimes seen on young trees, it almost always grows on old ones. Mycologists can therefore use it to identify old-growth oak and beech forests. It also grows on horse chestnut, black alder, hornbeam, aspen, and linden, often protruding from a crack or fissure in the trunk. In some European countries, we see it frequently; in others, it is increasingly rare because of environmental degradation and habitat loss.

Should we be concerned about the conservation of this species, given its growing popularity? Not if we buy cultivated hou tou gu or, since it's relatively easy, grow our own. Although the loss of old forests and woodlands presents a bigger threat to the fungus's survival than foraging does, hou tou gu is on Red Lists in eighteen countries and it seems like a good idea to just admire this beauty rather than pick it. Instead, enjoy your homegrown hou tou gu or perhaps buy the fungal fruits of those mushroom workers' early-morning labor.

6 AM

Fairy Ring Mushroom

Marasmius oreades

(WORLDWIDE)

Wandering into the garden at first light, we might blink twice as we discover a seemingly inexplicable phenomenon on the lawn.

Rings of mushrooms appear not just on lawns and in forests and fields but also in folklore, especially that of Western Europe. British and Irish fables dating back to the Middle Ages warn of the dangers of such "fairy rings": If you dare step into one, fairies may force you to dance until you collapse in madness or exhaustion; else all manner of other curses and ills may strike you down. The French and the German trade fairies for witches, calling the rings *ronds de sorcières* and *Hexenringe*, respectively. And if not the trap of a mythical figure, a ring of mushrooms may still mark the place of a lightning strike or shooting star, a surreptitious visit by a dragon or the devil, or even an unidentified flying object. I don't want to rob the mushrooms of their magic, but there's another story behind the fungus's growth in seductive circles.

Imagine a fungus spore germinating beneath the soil. It forms a mycelium, and as it feeds and spreads outward radially, the soil is depleted of nutrients. You might notice a paler patch in the grass above this action. The trailing parts of the mycelium die as food runs out, while the leading parts continue to break down organic matter with their enzymes, unlocking nutrients that result in lusher grass.

While there's food available, the mycelium

grows. But once the fungus exhausts the food supply or receives specific signals from the environment—such as colder soil—the mycelium switches from feeding mode to reproductive mode, shifting resources to produce mushrooms. Fairy rings form when a mycelium produces multiple mushrooms in a circular pattern. Sometimes a ring is incomplete, and the mushrooms appear in an arc, double arc, or sickle-shaped arc. If you're lucky enough to see a fairy ring pop up each year on your lawn, the widening aboveground formation mirrors the expanding underground mycelium. Depending on conditions, fairy rings grow about two inches a year, with the mushrooms appearing only on the outer edge of the circle, as the inner regions of the mycelium die off.

The fairy ring mushroom is not the only fungus to delight us with these enchanting formations. Another 140 species or so, including the field mushroom (*Agaricus campestris*), the blewitt (*Lepista nuda*), the fried chicken mushroom (*Lyophyllum decastes*), and many types of puffballs, also form fairy rings. While it's difficult to determine the age of fungi, we know that they can live to be very old. Some fungi that form fairy rings are estimated to have been around for at least two thousand years.

One of the largest fairy rings I have found was in the Black Forest in Germany—an auspicious discovery, considering how much theatrical folklore has arisen from this forest. After first checking for fairies, I followed the snaking arc of that trooping funnel (*Infundibulicybe geotropa*); it was a good hundred feet wide! Even so, my find was far surpassed by a ring in northeastern France that was formed by the same species but was ten times as big and estimated to be over seven hundred years old. Although the rings of the fairy ring mushroom are much smaller, they're every bit as entrancing.

The fairy ring mushroom's genus, *Marasmius*, was proposed for mushrooms that are marcescent—capable of reviving if they desiccate and shrivel up, allowing them to continue releasing spores. Few species have this remarkable adaptation for surviving dry periods. Most mushrooms are putrescent, meaning that they decay. Although they're small, fairy ring mushrooms are certainly not fragile. If you tug on the stipe of a fairy ring mushroom, you might be surprised to find that it's remarkably tough.

The fairy ring mushroom has attracted our attention with its magical rings, but it's also a popular edible species, especially in France. The French discard the tough stipes and cook the delicately flavored caps in butter and white wine. Beware, however, if you're foraging it for food, as it's one of many infamous and rather nondescript little brown mushrooms—or LBMs, among mycophiles. Be sure not to confuse this fungus with toxic look-alikes, such as fibrecaps (*Inocybe*) or poison pies (*Hebeloma*), that also grow in lawns. It can help to look at the specimen's spores, which are white in fairy ring mushrooms and black or brown in many of the look-alike species. But that alone isn't enough to rule out every poisonous doppelgänger, so do be careful. The fairies promise danger, even if you don't step in a ring!

7 AM
Porcino
Boletus edulis

(WORLDWIDE)

The clock strikes seven in the Piazza Galimberti, Cuneo's town square. The sun has barely risen. Sergio Ferrero leans on his walking stick, clucks his tongue, and barks, "Dai sbrigati!" Hurry up!

The Piedmont region in northwest Italy is famous not only for wine and white truffles but also for porcini. And for Sergio, porcino is numero uno. If you're not in the forest soon after sunrise, however, you're likely to come home with an empty basket. Only the early riser with keen eyes will reap the reward of savoring porcini.

Although some boletes—such as the devil's bolete we met at 3 AM—are toxic, many are

edible. Of these, the porcino is, without a doubt, the most revered. Few other fungi surpass it in flavor. It is known to Americans as the king, to Germans as *Steinpilz* (stone mushroom), and to the British as the penny bun. For Italians, *porcino* means "little pig." These common names reflect similar cultural interpretations of the mushroom's chubby form. Finns, on the other hand, cut to the chase and call the delicacy what it is: *herkkutatti*, "delicious mushroom."

Sergio and I head into the mountains, passing through woods of chestnut and larch, fir and spruce. The bladeless windshield wipers of his car squeal against the glass as the sky begins to drizzle. Sergio swears under his breath at the line of like-minded individuals winding their way up the mountain. It is late September, and decent rain in previous weeks has improved our chances of finding porcini—that is, so long as we beat the others to them.

Mushroomers keep their foraging spots secret, but porcini are not so difficult to find.

For a start, they're big, reaching the size of a dinner plate. If you're planning on eating them, however, the biggest ones are not what you're after, as maggots and other spineless creatures are likely to find them before you do. Like many boletes, porcino has a bulky form. Its cap can be various shades from light brown to reddish brown, with a slightly greasy or tacky texture. The firm pore surface on the underside starts out white, becoming yellow as the fungus matures. Covering the bulbous stipe is a fine, white netting, which can be quite subtle and require a close look.

Although porcino's distinctive form and color make it easy to recognize, foragers still sometimes confuse it with toxic look-alike species. In the early nineteenth century, foraging mushrooms both for personal use and to trade was common—as were cases of poisoning. In 1820, in an attempt to reduce the incidence of poisoning, the Austro-Hungarian Empire instituted the first set of rules for wild fungus consumption and commerce. Other regions with foraging traditions, such as northern Italy (then part of the Kingdom of Sardinia) followed suit, introducing their own laws for mushroom collection. Today, many regulations also have a conservation imperative to prevent overharvesting and environmental harm.

Across the northern Italian border in Switzerland, foragers have enjoyed the certainty of mushroom identification with the expert help of *Pilzkontrolleure*, or "mushroom inspectors," for over a century. My Swiss friend Barbara Thüler is a mushroom inspector and, like Sergio, has had a lifelong love affair with porcini. When Barbara is identifying a mushroom, she uses all her senses to glean as many clues as possible about its identity. She'll examine it carefully, especially its underside, then close her eyes and hold it to her nose. Fungi have extraordinary scents and aromas, as you'll discover throughout this mushroom day, and a good sense of smell makes mushroom identification easier.

The more you know about a fungus species, such as where it grows and the trees with which it associates, the less likely you are to mistake it for another. But the savvy forager learns to identify not only the edible fungi but their toxic doppelgängers too. In time—and indeed it takes time to gain diagnostic experience—foragers get better at anticipating when and where to find a particular fungus. Thorough knowledge of just a handful of mushrooms is safer than superficial knowledge of many.

Sergio has been foraging porcino since childhood. He was taught by his father, who was taught by his in turn. After a few hours in the woods, Sergio looks at the basket brimming with porcini, plucks a stray strand of tobacco from his mustache, and slowly nods. I take it as a sign that he is happy with the day's haul. In my excitement to get the porcini home and into the pan, I speed down the forest path, as Sergio bellows from behind, "Piano piano!" Slow down!

8 AM
Chi-Ngulu-Ngulu
Termitomyces titanicus

(AFRICA)

If your eyes aren't quite open by 8 AM, don't worry. You'll have no problem spotting the termite mushroom, or chi-ngulu-ngulu. It's not just termites that are hunting this fungus; human beings are too.

The edible chi-ngulu-ngulu grows in the Miombo woodlands of south-central Africa. It's known for its smoky flavor and meaty texture, and the great thing is that you only need one to feed the entire family! As its specific epithet, *titanicus*, suggests, it's the world's largest mush-

room, with its cap reaching the spacious span of an umbrella. This fungus is collected toward the end of the rainy season (October–November), often by women and girls, who sell it at roadside stalls and markets. Local people eat chi-ngulu-ngulu fresh, but they also dry it out so that it can be ground into a powder, then added to sauces and stews.

Now look at the genus name, *Termitomyces*. This mushroom is not only impressively large; it has a fascinating backstory of an intimate alliance with other organisms: termites—the small, soft-bodied and social insects that feed on wood and other decaying plant matter.

The life cycles of fungi and termites have aligned for thirty million years, possibly longer. Some biologists describe their relationship as agriculture, with the termite farmers tending their fungus crop. But unlike humans, who farm for the crop, these termites exploit the fungus for the decomposition it provides.

Most termites use microbes in their guts to

digest the lignocelluloses that give plants their hardness. Fungus-farming termites have teamed up with chi-ngulu-ngulu. Different insect species approach farming differently, but this unlikely union seems to have mastered circular agriculture. How does it work? The termites consume fresh plants and then build combs—cultivation chambers that look a little like brains or sponges—out of the partially digested plant material in their fecal pellets. Next, they collect and deposit chi-ngulu-ngulu spores in the combs. When the spores germinate, the resultant mycelium consumes the combs, producing a rich compost packed with sugary goodies on which the termites feed. The termites also feed on the fungus itself. Spores that develop in tiny nodules on the fungus hyphae survive the journey through the termites' guts and, once excreted, form new combs. There could be as much as forty kilograms (ninety pounds) of fungal combs within a mature colony. That's about the weight of four large watermelons.

Chi-ngulu-ngulu and the termite rely on each other for survival. The fungus benefits from a protective termite mound, allowing it to live in environments that are otherwise too dry. And the termites benefit from the fungus's help with digesting plant lignocellulose. There's a third winner in this relationship as well—the environment. Together, the fungus and insect contribute to plant decay, improving soil composition and fertility.

The fungus not only prefers to live with termites, but it seems incapable of living without them. When researchers tested this theory by removing the combs from termite mounds, other fungi (mostly species of *Xylaria*) moved in and outcompeted chi-ngulu-ngulu. The termites are fastidious in keeping their combs clean and free of other fungus competitors, ensuring the maximum growth of chi-ngulu-ngulu.

When the rains arrive, chi-ngulu-ngulu produces mushrooms. Because the fungus is within the termite mound far underground, its mycelium needs a few special tricks to get the mushrooms above the soil's surface. The first is to grow a long stipe—about the height of an adult's knee—called a pseudorhiza, which rises from the fungal comb. The second is to develop a cap with a special knob-like perforatorium that can pierce the mound and penetrate the hard soil.

Like many fungi, chi-ngulu-ngulu can reproduce sexually through the spores formed on its mushrooms and asexually through the conidiospores formed directly on the mycelium. Asexual reproduction allows the fungus to spread within the termite mound, while sexual reproduction enables it to disperse beyond the mound and colonize new host nests.

Chi-ngulu-ngulu and its colonies of fungus-farming termites have formed a marvelous mutualism that allows them to disperse widely and colonize inhospitable environments. It is one of the many survival strategies that fungi have evolved, but the big question remains: How did this unlikely union come to be? There are various theories, and the ingenious innovation of the comb is a likely clue. Let's just hope that it won't take mycologists thirty million years to find the answer!

Veiled Polypore
Cryptoporus volvatus

(ASIA AND NORTH AMERICA)

Listen carefully. Do you hear that morning tapping? It's a white-headed woodpecker (*Picoides albolarvatus*) hunting for a breakfast of bark beetles. These avian excavators are essential to forest ecosystems. Their job is to create cavities, but they don't do it alone: The veiled polypore first softens the wood with its enzymes. The fungus is itself home to bark beetles and other insects on which woodpeckers feed, and foraging birds help spread spores far and wide. These entangled associations among fungus, tree, bird, and beetle are just some of the many complex relationships that underpin functioning forests.

The veiled polypore can be found through-out North America and Japan, and mushroom chasers have also spotted it in South America and eastern Asia. It grows on conifers, usu-ally pines that have died but are still standing. Although the trees are dead and can no longer photosynthesize—that is, harvest energy from the sun—they have intricate afterlives as hosts for a diverse succession of organisms, including fungi. These organisms play vital roles in pedo-genesis, converting wood to soil. Indeed, the veiled polypore is a pioneer fungus species: it kick-starts recycling by secreting enzymes, then colonizes the sapwood and accelerates decay.

The veiled polypore is a fungal oddity, looking a little like a small puffball stuck on the side of a tree. Unlike other polypores (hard hoof-shaped fungi with pores on their undersides), it sports a volva-like flap of tissue, or a veil, that conceals its fertile pore surface, creating an enclosed pouch or chamber. The veiled polypore's generic and specific names reflect this anomaly: *Cryp-*

toporus means "hidden pores," and *volvatus* means "with a volva."

But why does the veiled polypore have a veil? If you look closely, you'll see tiny clues in the form of nearby boreholes that penetrate the tree's inner sapwood. Mycologists think that the veiled polypore harbors wood-boring beetles, with the veil not only maintaining the ideal moisture and temperature conditions for the inhabitants but also ensuring that they are decently dusted with spores. Given that this fungus releases its spores during the drier months, providing a favorable home for its spore-dispersing helpers is crucial.

As the veiled polypore matures and its spores develop, a hole at the base of the veil begins to enlarge. Wood-boring beetles, along with other insects, enter the moist and nutritious chamber via this convenient doorway. There, they feed on the hymenium and spores, and some even use the protective pouch to pupate. If you slice a veiled polypore open, you'll see layers of tubes and flesh beneath its veil, not to mention the creatures and spores that are often crammed inside its chamber. Once the insects have fed and bred, they exit through the same doorway and fly off, ferrying spores to new habitats. Any spores that fall out of the hole are whipped away by the wind. It's an ingenious way to ensure two modes of spore dispersal. But wait, we've already mentioned a third!

The white-headed woodpecker also contributes to spore dispersal by foraging for beetles within the fungus. We know that insects eat and distribute spores, but most insects don't travel too far. Imagine how much farther spores could get if they hitched a ride with a bird, especially a migratory bird that might traverse continents. As the woodpecker feasts, it dislodges spores and hypha fragments, which then accumulate in its plumage. When the bird flies off, it transports this fungal cargo from one tree hollow to another, allowing the fungus to colonize new habitats. The bird not only gets a meal and helps the fungus reproduce but promotes the decay of other dead trees, increasing habitats for saproxylic (wood-reliant) organisms.

Cooperation is the foundation of life. As we have seen, symbioses often entail complex interactions between multiple species and benefit not just those that are directly involved but others in the ecosystem too. As forests change, however, older hollow-forming trees are vanishing. Although artificial hollows and birdhouses are well-intentioned and offer an alternative home for some species, they do not replace the complexity of the natural environment. It's not just woodpeckers that rely on hollows but numerous birds and myriad other organisms, reminding us how vital fungi are in contributing to forest function and diversity.

Lawyer's Wig

Coprinus comatus

(WORLDWIDE)

Also known as the shaggy mane, ink cap, or shaggy ink cap, the lawyer's wig offers us a fungal disappearing act. In a process called autodigestion (or autolysis), this species "eats" itself as it decomposes, metamorphosing from a slender, fluffy white mushroom into an inky, black mess.

Lawyer's wigs are so called because of the unusual, tufted texture of their cylindrical caps, which resemble the traditional wigs worn by barristers and judges. Another common name, ink cap, refers to its black spores and the slurry of "ink" that results from autodigestion. Why autodigest? It's an ingenious adaptation for dispersing spores. Unlike those of most mushrooms, the lamellae of lawyer's wigs are so close together that there's little room for spores to fall out. A bit of fungal dexterity solves the problem. As its spores mature, the cap curls outward, increasing the space between the lower edges of the lamellae. Presto chango! Some spores escape and float off on the wind. The wig disintegrates as maturing spores are released from the bottom to the top. This fungal Houdini—once tough enough to push through asphalt—now collapses into a formless ooze.

The lawyer's wig digests itself using enzymes that break down chitin, the compound that gives strength and sturdiness to fungus cell walls. Carefully timing the autodigestion is critical: The fungus must produce chitinase in its cap and lamellae immediately before releasing spores. Once the mature spores are dispersed, this potent digestive dissolves the cells of the lamellae up to the point of maturing spores.

But the lawyer's wig digests more than itself.

Beneath the soil its mycelium has a surreptitious strategy for stealing a little extra nutrition—it strangles wriggling nematodes. A voracious predator, this fungus specializes in trapping, killing, and digesting those worms. Nemato-phagous fungi, a diverse group that includes the lawyer's wig, poach nematodes for their nutrients, especially nitrogen. First observed in the 1800s, this form of hunting appears to have evolved independently among many different fungus species. It's a clever means to access more food, paralleling the way insectivorous plants absorb nitrogen from spineless prey trapped in their modified leaves.

We know that over seven hundred fungus species digest nematodes, and they've developed a suite of deadly devices to immobilize their prey. The lawyer's wig possesses potent toxins and tiny spiny balls at the ends of their hyphal branches. These lethal weapons can break the nematode's equivalent of skin, causing the animal's internal fluids to leak out. The fungus hyphae then penetrate the nematode via its wounds and consume its innards. It's swift death for the nematode and fast food for the fungus.

Other fungal nematode killers use hyphal trapping devices, such as sticky nets and knobs, constrictive rings, and glue traps. Some fungi can even wrangle their hyphae into lassos. If a nematode enters the loop, the fungus contracts, strangling its prey. Not all fungi mount external attacks—some strike from within. Endopara-sites, for example, live within the nematodes, entering via their mouths or penetrating their skin, then consuming them from the inside out. Other fungi have sophisticated chemical surveil-lance systems, detecting molecules that nema-todes secrete into the soil. If the fungus is feeling hungry, it lays a snare. Then there are those fungi, such as oysters (Pleurotaceae), that not only trap nematodes but also give them a nasty dose of toxin just to be sure they don't escape before they're devoured.

Many humans, in turn, like to digest this

tasty edible species. Of course, you'll want to eat it while it's young, before it digests itself. You'll have only hours after picking it before the wig dissolves into an inky slop. Be sure to get it in the pan fast but only after you've checked that you haven't confused it with the inky cap (*Coprinopsis atramentarius*), a look-alike we'll meet at 9 PM.

Golden Chanterelle

Cantharellus cibarius

(NORTHERN HEMISPHERE)

By 11 A M, the iconic Östermalms Saluhall in Stockholm is bustling with buyers of fresh produce and foraged mushrooms. As I meander down the market aisles, I see vendors pile their stands high with orange fungal delights. Foragers have warned me on more than one occasion never to get between a Swede and her chanterelles, so I'm not taking any chances. Yet, digging deeper into Scandinavian history, it seems the Swedish obsession with the golden chanterelle is relatively recent.

In many European countries, the rural poor foraged for mushrooms as a subsistence or, in times of extreme hardship, famine food; in some countries, they still do. In Sweden, however, the history of mushroom foraging took a different path. It was the urban elite who developed a taste for fungi. Indeed, the first written record of edible mushroom consumption was from the dinner table of the Royal Palace of Stockholm on July 10, 1636.

Anders Hirell is a historian who studies Swedish foraging. I met Anders at a mycological meeting in Uppsala and asked him why the fervor for fungi unfolded differently in his homeland. Attitudes toward fungi vary across cultures, and Swedes did not initially regard mushrooms as food. "They would eat grass, straw, bark, roots, leather boots, sawdust, even horse excrement," he said, "but not mushrooms." Then something changed.

In 1818, in a strange twist of history, Sweden found itself with a French king, Karl XIV

Johan, who brought with him his love of French cuisine—including *champignons*. Alongside the influence of royalty, scientific interest in fungi was also growing. Although the famous Swedish biologist Carolus Linnaeus, who formalized the system for naming organisms known as binomial nomenclature, had only minimal interest in fungi, another Swede, Elias Magnus Fries, was not only laying the foundations of fungal taxonomy but was an avid fan of fungal fare. In 1860 the Swedish Royal Academy of Sciences commissioned Fries to produce Sweden's first book on edible mushrooms. Yet despite Fries's book and the efforts of mycologists to teach people about edible mushrooms, it took another century for fungus foraging to become popular.

Identifying fungi is a tricky business, but with few toxic look-alikes, the golden chanterelle is among the easiest edible mushrooms to recognize. With its unusual form, conspicuous color, sweet fruity scent, and tendency to appear in great abundance, this species slowly shifted

Swedish attitudes toward fungi from disdain to reverence.

Regulations and guidelines around foraging vary by country. Many European countries impose restrictions on the quantity and species that foragers can collect. But in Sweden, *Allemansrätt*, or "all people's right," gives everyone the freedom to roam in nature and forage for mushrooms and berries. Most Swedish foragers find fungi for pleasure and recreation, not for survival.

Today, fungi have finally been freed from the stigma of famine food, and the chanterelle occupies a special place in Swedish cultural identity. Golden chanterelles have enamored not only Swedes but other Europeans too. Of the 268 types of mushrooms that are commercially available across twenty-seven European countries, only the golden chanterelle and the porcino are sold in every one, attesting to their popularity.

There are fewer than eleven million people in Sweden, but on a balmy fall afternoon, it can feel like half the population is alongside you in the Swedish forests, hunting for chanterelles. Could the intensity of foraging threaten the future existence of the Swedes' favorite fungus? American mycologist Lorelei Norvell conducted a decade-long study in the forests of America's Pacific Northwest to determine the effect of foraging on golden chanterelle populations. The study revealed that harvesting does not affect the chanterelle's ability to produce mushrooms, just as plucking pears from a pear tree shouldn't hinder it from bearing fruit in the next season as long as the tree is not damaged in the process. A longer-term study of multiple fungus species in Switzerland also found that mushroom harvesting did not affect future yields.

But there may still be cause for concern. The Swiss study showed that harvesting disturbs the forest floor—and the fungus mycelia below. Too many harvesters can compact soil, alter water flows, introduce pathogens, and dislodge the organic matter that provides fungi with moisture and food. While studies are helpful, they have their limitations. Researchers cannot always apply findings from a study of one fungus species to predict how harvesting will affect others. Understanding the complexities of how fungi operate in forests requires nuanced knowledge of their preferences for moisture levels, nutrient availability, and the type of organic matter in which they live.

As the groundswell of interest in fungus foraging continues, a cautious approach may minimize harm and ensure that we can dine on the sweet and sumptuous chanterelle for a long time to come.

Arctic Orange-bush Lichen

Seirophora aurantiaca

(CANADIAN ARCTIC)

By midday, the thick fog lifts, allowing weak rays of sunshine to penetrate the wild and windswept coastline. Few organisms can tolerate the harsh conditions of Canada's Western Arctic, but one superb strategy for surviving extremes is to team up with others. Lichen-forming fungi have had a few hundred million years to perfect their relationships and can colonize some of the most inhospitable habitats on earth, from blistering desert and bare, wind-blasted rocks to tundra and toxic slag heaps.

The Arctic orangebush lichen ekes out an existence among the rocky cracks and crevices of ice- and wind-scoured shorelines. It is known to occur in only twelve scattered spots in the Inuvialuit Settlement Region on Banks, Melville, and Victoria Islands and the Cape Parry area of the Northwest Territories mainland. If it can't find a suitable crack, it'll settle for some rocky, hummocky tundra or an old beach ridge close to the coast. The midday sun is vital to the survival of this fungus because it does something that other fungi don't: It photosynthesizes, convert- ing sunlight into sugar. Well, at least it's part of a pioneering partnership that does.

It can be tough doing it alone, and lichens understand this. The lichen lifestyle involves a liaison between a fungus and a photosynthetic organism—either a green alga or a specialized blue-green alga called a cyanobacterium—that taps the energy of the sun. These fungus-alga duos double their talents, with the fungus providing a thallus, or protective body, within which the algal cells nestle and turn sunshine into food. They thus increase their resilience and chance of survival.

The lichen entity is made up of additional players with diverse ecological roles. It comprises single-celled fungi called yeasts, as well as bacteria, viruses, and protists. Lichenologists and other scientists who study lichens recently discovered these stealthy collaborators while examining lichen DNA. They believe yeasts help lichens defend themselves, warding off would-be predators with a cache of distasteful chemicals. Bacteria could contribute to several essential functions, helping the lichen meet its nutritional needs by supplying nutrients, especially nitrogen, phosphorus, and sulfur. They may also play a defensive role against pathogens and support photosynthesis by providing vitamin B12. Protists contribute by moving nitrogen to where it's needed in the lichen's growing parts and controlling bacterial populations. Although researchers have studied lichens for over two centuries, the complex give-and-take of inter-actions within the lichen microcosm remains elusive.

Lichenologists refer to the Arctic orangebush lichen as fruticose because it has a shrubby or bushy structure. Other lichens may be foliose (flatter with an upper and lower side), crustose (even flatter and forming a crust), or leprose (powdery). Although the Arctic orangebush lichen is small, its bright, orange color is conspic-uous. Given that the species is nonetheless hard to find, scientists have concluded that it's rare.

The special pigments that give lichens their orange, yellow, or red coloration are called anthraquinones. They help lichens survive in extreme environments by providing protection against damaging ultraviolet light, but they're often poisonous to the lichens. Thankfully a spe-cial transporter gene keeps these pigments out of the lichens' cells, allowing the intense orange hues to reflect sunlight without causing toxic effects.

While the Arctic orangebush lichen has worked out how to occupy the margins of life, it faces a bigger threat beyond its control. Its survival hinges on a fragile habitat that is exposed to the consequences of the changing climate in the Beaufort Sea, especially the altering weather patterns and diminishing sea ice. Increased ice melt and saline wash from storm surges, as well as melting permafrost, reduce its habitat. As the climate warms and the summer extends, plant communities change. The encroachment of vegetation and invasive species could also threaten the lichen's existence.

Climate change operates on a global scale, but its effects are more dramatic in the Arctic. Temperatures have increased there at about twice the rate as elsewhere. As the bright and reflective ice melts, it exposes the darker ocean below. Water absorbs more of the sun's heat than ice does and therefore amplifies the warming. Coastal environments are prone to sea-level rise, erosion, and permafrost melt because they're almost entirely made up of icebound soils, ground ice, and unstable sediments. In the summer, soils thaw more deeply into the earth. Erosion along shorelines exposes permafrost, which increases melting. This can cause landslides known as retrogressive thaw slumps, which occur over vast areas.

Scientists think that the Arctic orangebush lichen population will decrease over the next century because of rising sea levels and the erosion of coastlines. But help is on the way. The research of lichen expert R. Troy McMullin and botanist Paul Sokoloff led to the inclusion of the Arctic orangebush lichen on the IUCN Red List in 2020. Researchers plan to visit areas where the species has previously been found to check whether it is still present and to search for new sites where it might grow. The Arctic orangebush lichen is not only a hardy extremophile but an ambassador for interkingdom cooperation and a reminder that teamwork makes the dream work.

1 P M

Penicillium

Penicillium roqueforti

(WORLDWIDE)

Who's hungry? After a morning hiking through the verdant hills and valleys of Aveyron in southern France, it's time to stop for lunch. From creamy cambozola to beautiful bleu de Bresse, the French are famous for their great range of delicious cheeses, thanks to their long-standing partnership with the microfungus, or mold, known as *Penicillium*.

One of many species of *Penicillium*, *P. roqueforti* is an important nutrient-recycling fungus found in soil, wood, and other organic matter in damp and cool environments. Because it can survive extreme temperatures and low oxygen levels and is resistant to many preservatives, it also turns up in places we don't want it, spoiling our dairy, wheat, and meat. Some molds are harmful; others are useful; and still others, like *Penicillium*, can be both. Although it degrades some foods, we wouldn't enjoy the delights of blue-veined cheeses, such as Roquefort, Stilton, and Gorgonzola, without it.

Cheese making is an ancient craft that began about 2,600 years ago. Today, over one thousand types of cheese exist, each with its own signature characteristics. Those unique tastes and aromas come from the way in which they're made, including how microorganisms—like molds, yeasts, and bacteria—turn liquid milk into solid curds. Cheese makers use *P. roqueforti* as a starter culture to produce various types of blue cheese. It creates those distinctive blue veins, or interior marbling, and gives the food its characteristic tangy punch and crumbly yet creamy texture. Taste and aroma are further enhanced as the fungus contributes to the breakdown of the milk's proteins and fats. It's astonishing that mold can transform ordinary old curds into treasured delicacies.

Different blue-veined cheeses have different organoleptic properties, or sensory characteristics, including aroma, flavor, appearance, and texture. The specific strains of *P. roqueforti*, type of milk, and manufacturing methods used, along with the environmental conditions, cellar temperatures, and ripening times, shape these qualities. Producers use only raw (unpasteurized) sheep milk to make Roquefort, and not the milk of any old sheep but exclusively that of the Lacaune breed from southern France. These hardy and adaptable sheep produce milk with high levels of fat and protein, which contribute to the cheese's intense flavor. Grazing in legume-grass pastures ensures the high volume and nutritional richness of Lacaune milk.

Food production is tightly linked to place and cultural identity. Cheese makers have passed down recipes for generations and protect them fiercely. The tiny village of Roquefort-sur-Soulzon sits among dramatic cliffs and the rugged terrain of the Causse du Larzac. It's not only the panoramic vistas that draw visitors to the town, however, but the clandestine activity deep in the damp underworld network of limestone caves below. Created by a series of ancient landslides and thousands of years of erosion, the caves' consistent temperature and humidity provide the perfect microclimate to foster the growth of *P. roqueforti* and the maturation of cheese.

The French verb *affiner* means "to refine." Cheese affinage is the art of aging cheese. The *affineur*, or "cheese maturer," determines when a cheese has reached its desired flavor and texture. Affineurs must carefully nurture the cheese during the ripening period because *P. roqueforti* is a fussy fungus. Slight changes in temperature or humidity can alter its growth rate, enhancing or ruining a cheese's quality. Affineurs control airflow with doors attached to *fleurines* (small natural openings in caves) and thereby regulate the temperature and humidity to suit the needs of the fungus.

Roquefort has been produced since at least the eleventh century, making it one of the oldest types of blue cheese still in production. Its ripening period in the caves is pivotal to yielding the desired characteristics and qualities. France imposes strict standards on the production of certain agricultural foods. To label a cheese *Roquefort*, makers must ripen it in the caves for at least three months, although many choose to age their cheese for longer to maximize the depth and complexity of its flavor. While blue cheeses are produced elsewhere in the world, only those that ripen in the caves can be called *Roquefort*—in the same way that not all sparkling wine is champagne. Selling "Roquefort" from anywhere other than Roquefort-sur-Soulzon is not just a sham but a punishable crime.

In the past, cheese makers would collect *P. roqueforti* from rotten bread in the caves; now, they produce the fungus in the laboratory, giving them more control and thus more consistent cheeses. Selecting for favorable traits to meet the strict specifications, however, has reduced the genetic diversity of the species. Although *P. roqueforti* can reproduce both sexually and asexually, the cheese-making industry has selected for asexual strains, cloning a single line to allow more predictable cheese production. No new genes are introduced in this process. Over decades, that strain has lost its vigor and accumulated genetic defects that have weakened it almost to the point of infertility. Until recently, there were only four populations of *P. roqueforti* worldwide, and of the two that are used in cheese making, only one is to produce Roquefort.

Fortunately, researchers have sequenced the genome of a previously unknown population of *P. roqueforti* to create a little-known cheese from the French Alps called bleu de Termignon. It just might be the lifeline that revives the fungus's genetic diversity and saves blue cheeses from extinction.

Branched Shanklet

Dendrocollybia racemosa

(EUROPE AND NORTH AMERICA)

It is early afternoon in the woods of eastern Canada, and fungus enthusiasts Sarah Riley and Josh Wayborn are poking around hemlock and cedar. For years they've been hunting a very special fungus, and it turns out to be their lucky day. There, beneath the sword ferns and salmonberry, they discover an enigmatic and seldom seen mushroom—the branched shanklet.

Many fungi have peculiar forms, but the branched shanklet's is a leading contender for most bizarre. Not only does the species have an

unusual appearance, but it has an extraordinary lifestyle as well; it feeds on the corpses of other mushrooms. Josh and Sarah had hunted the shanklet for years, checking among mummified mushrooms on every foray. When at last they spotted its discreet yet distinctive form, they recognized it straight away.

Compared with the fairy ring mushroom we encountered at dawn, the shanklet is a little more fragile. Its tiny caps rarely reach the size of a dime. They start out curved, flattening at maturity, and often have a raised central knob known as an umbo. Oddly, its caps are sometimes absent altogether.

The more unusual part of its anatomy is its stipe. It's adorned with short lateral outgrowths, or side branches, that are equipped with structures called slimeheads (which would be a superb name for a punk band). The fungus packs these swollen tips of its branches with asexual spores, while its lamellae produce sexual spores. Since the slimeheads are wet, invertebrates or even small mammals that come up against them probably help disperse the spores. But no one has observed this happening, so we don't know for sure. Humans don't always appreciate organisms whose appearance bucks the trend. Mycologist George Massee was not so enamored with the branched shanklet, describing it as a "monstrous or abnormal form of some species." Through another lens, preferred perhaps by the many humans who themselves long to be unusual, this unorthodox yet elegant fungus has nailed it!

The stipe of the shanklet emerges from a black egg-shaped ball of nutrients, or sclerotium, that helps the fungus survive harsh winter or summer conditions. The sclerotium grows from deep within the duff—of conifer needles or, more often, other decomposing mushrooms. Because the host mushrooms are often decayed, it's difficult to identify them, but mycologists think they're milkcaps and brittlegills. The shanklet might opt for the brittlegill species

Russula crassotunicata, which reputedly smells like coconuts, because it appears in large quantities for many months, offering a reliable growing medium and long-lasting food source. If it were me, the coconut odor would also be a selling point.

Mushrooms that grow on other mushrooms aren't common, but a few other fungi took this evolutionary path too. Some grow on a living host, while others seem partial to one that's putrefying. Mycologists have observed true parasitism in some of these relationships—such as that of the lobster mushroom (*Hypomyces lactifluorum*) and its host, which it renders infertile—but not in others. The shanklet, for instance, simply finds nutrition in its decomposing hosts.

Although the branched shanklet is widespread, it's locally rare, appearing on Red Lists in the United Kingdom, Norway, Denmark, and the United States. Determining where a species lives is the first step toward its conservation. Historically, plants and animals have been the focus of biodiversity surveys, but fungi are finding their way into biomonitoring and conservation programs. The branched shanklet is one of the target species of the Fungal Diversity Survey, a project aimed at documenting, understanding, and protecting fungi in North America. Other fungus-monitoring initiatives, such as the Lost and Found Fungi project in the United Kingdom and the Fungimap project in Australia, bring together professional mycologists and the wider community. Mycophiles like Sarah and Josh contribute to our understanding of the distribution and conservation status of the many amazing—and idiosyncratic—members of the fungal kingdom.

Sandy Stiltball

Battarrea phalloides

(WORLDWIDE)

It is midafternoon in midsummer in Portugal's
Monsanto Forest Park, and the heat is severe.
The last thing I expected to encounter among
the wilting barley grass was a fungus. Yet there
stands an old friend, the sandy stiltball, on its
shaggy stipe. Also known as the scaly-stalked
puffball or desert-stalked puffball, the species is
impressive for its stature, growing almost to the
height of my knee. Although tall, it was difficult
to spot because it blended in so well with the
beige and brown tones of its dry surrounds.

How can you recognize it? The sandy stiltball
is like a puffball on a stick. For many mush-

rooms, spore release involves both an active and a passive phase. In the active phase, a mushroom uses a surface-tension catapult to explosively eject its spores—a phenomenon called ballisto-spory. The wind then takes over, carrying them to new habitats. Puffballs, unable to forcibly discharge their spores, have had to come up with alternatives. Some puffballs release spores from small openings, relying on an insect or the pressure of a raindrop to push the ball's surface down enough to force the spores upward and out through the hole. Others play a numbers game. The basketball-sized giant puffball (*Calvatia gigantea*), for example, produces a whopping seven trillion spores, hundreds of times more than your average mushroom.

Most puffballs seem content to sit on the ground; these are known as "true" puffballs. The sandy stiltball and other stalked puffballs, however, take to the air—not for the view but to gain a height advantage in spore release. The ball at the end of the stick is a mass of powdery spores, or gleba, enclosed with a protective layer called a peridium. The peridium itself has outer and inner layers. As the immature egg-like structure of the sandy stiltball develops beneath the soil, the stipe pushes through the outer peridium, leaving its remnants at the base as a membranous cup, or volva. The stipe then extends upward, hoisting the gleba into the air. At maturity, the inner peridium ruptures along its margins and falls away, exposing a rusty-brown mass of spores to the wind.

The sandy stiltball doesn't grow only in Portugal but in dozens of other countries, scattered across every continent except Antarctica. Despite its global distribution, it's seldom recorded and is considered rare in some countries. The sandy stiltball grows in dry and, well, sandy habitats—from steppes and savannas to deserts, coastal dunes, and woodlands. With its tough woody stipe, it can withstand the hostile conditions of desert existence, such as extreme temperatures and abrasive, sand-laden winds. It grows in other sorts of environments too. It has been found in subtropical rainforests in Brazil, on alluvial riverbanks in Bulgaria, and in a juniper forest in Macedonia. The sandy stiltball also appears in ruderal environments (those that have been disturbed by humans), from old dunghills and sawdust heaps to graveyards and hedgerows.

Populations of the sandy stiltball appear to be stable in some countries and threatened in others. It has disappeared altogether from Serbia and Poland. It's hard to determine a species' risk of extinction, or even whether it's rare or just rarely observed. (Recall the Arctic orangebush we met at noon.) The changing whereabouts of fungi raise interesting questions about rarity, especially when a species deemed rare flourishes in human-disturbed habitats. But how much disturbance is too much? As some countries convert stiltball habitats into parking lots and housing complexes, the fungus could well disappear from them too. Or perhaps it'll take advantage of those disturbances to claim territory that is uninhabitable by other species.

The hardiness of this fungus despite the degradation and loss of many of its natural habitats gives me hope. The sandy stiltball and other stalked puffballs are a tough bunch. And although not poisonous, they're not edible either; you might break a tooth if you try to bite into one. The sandy stiltball can persist long after it has dried out and its spores have blown away. I've been amazed to see the spike of one's stipe still standing tall in the Australian desert a year after I first found it.

As the temperature climbs yet higher in Portugal, I start to flag. I tell the stoic sandy stiltball that I'll come back to visit next year, then I head off in search of a cooling ice cream.

Aniseed Funnel

Clitocybe odora

By late afternoon on a mild fall day, we're in the French Jura. An unexpected aroma wafts through the woods. Could it be licorice? aniseed? maybe even ouzo? The odors are similar, but this one is actually coming from a fungus: the aniseed funnel.

Appearing in stunning shades of teal and turquoise, the aniseed funnel looks every bit as enticing as it smells. Because its colors are unusual in fungi, there are few species with which it can be confused. The verdigris agaric (*Stropharia aeruginosa*) shares a similar hue but lacks the unmistakable scent of the aniseed

funnel. French mycologist Jean Baptiste François Pierre Bulliard considered its odor such a defining feature of the fungus that he chose the specific epithet *odora*, meaning "perfumed."

To understand any mushroom well, it helps to observe it at different developmental stages. By looking at all its parts, you become familiar with how they vary and change. Like most mushrooms, the aniseed funnel is umbrella-shaped. Its cap starts out convex (rounded outward), with the margins inrolled (tucked under). It often has a delicate sprinkling of white dots. These are the membranous remnants of a veil that joined the cap to the stipe in its immature stage and protected the developing spores. As the mushroom grows, the cap margins unroll and become wavy and the center sometimes rises into an umbo. The cap then flattens out, its edges lifting and its center dropping, until it becomes concave, like a shallow funnel. But alas, with age, its magnificent blue-green tones fade to gray and its alluring aroma subsides.

The aniseed funnel grows across temperate zones in Asia, Europe, and North America. It's a saprotrophic, or recycling, fungus. You'll find it in both conifer and broad-leaved woods, but you're more likely to smell it before you spot it. Its odor comes from a volatile compound called p-Anisaldehyde. Some plants, including anise and Korean mint, also contain this aromatic chemical, and the flavor and fragrance industries use a synthetic version. While humans respond to this agreeable aroma, its benefit to the fungus is unknown. Perhaps smelling so sweet also attracts insects that spread its spores.

If you're fortunate enough to forage with some European elders, you'll notice how they almost always smell a fungus to identify it. Today, many people in Western cultures are less skilled at recognizing certain odors and mask those they deem disagreeable. Humans' intolerance of odors considered "off," and the resultant artillery of products to disguise them, has reduced our reliance on this remarkable sense.

Sensitivity to smell depends on age, health, life experience, whether we're using medication, and whether we're pregnant. Impressions of fungus odors vary more than those of features like color, form, or texture. Not only is sense of smell subjective, but we often find it hard to describe the odors we can detect. In English, we have a limited olfactory vocabulary and often resort to vague or blunt descriptions of fungus odors, such as "mushroomy," "earthy," or "unpleasant." Although many such odors are indistinct to a human nose (hence why dogs and pigs are used to hunt some fungi, like truffles), it's generally easier to detect fungus scents when the air is warm. So, if you can't get a good whiff of a mushroom, wait for the warmest part of the day or try popping it in your pocket to warm it up a little. Or, if that feels a bit too icky, enclose it in a small container—that should do the trick.

Few fungi smell like the aniseed funnel; the aniseed cockleshell (*Lentinellus cochleatus*) and the abruptly-bulbous agaricus (*Agaricus abrupt-*

ibulbus) are two species that do. Fungus odors can resemble a great range of foods beyond aniseed, including garlic, radish, potatoes, fenugreek or curry, apricots or ripe pears, honey, burnt sugar, and even chocolate or bubblegum. Other fungi have a chemical edge and smell like antiseptic (iodine), bleach (chlorine), moth balls (naphthalene), or rotten eggs (sulfur).

While the aniseed funnel's odor is unequivocal, opinions about its edibility are mixed. Some people suggest that it could be poisonous and warn against eating it, while others say that it's safe if unpopular. When foraging for any fungus, however, you want to be one hundred percent sure that you can differentiate it from toxic look-alike and smell-alike species. If you're unsure, it's better to simply turn up your nose.

Anemone Stinkhorn

Aseroë rubra

I hope you got a good whiff of the last perfumed fungus we met. Now, you might want to hold your breath. In the late afternoon in a sun-warmed Australian forest, eucalypts release their heady mix of oils and resin as the sun slides toward the horizon. But whoa! What's that wretched stench?

The anemone stinkhorn is deserving of its name. But before we take a deeper sniff, let's start with its appearance. This fungus looks like

something from a Tim Burton film. Its form differs from that of other so-called phalloids, appearing more like an anemone or starfish. The more predictably phallus-shaped members of the Phallales order may elicit chuckles or gasps of horror from passersby, while the eccentric anemone stinkhorn is more often met with furrowed brows. Whereas an umbrella-shaped mushroom commonly comes to mind when people imagine a fungus, this striking beauty sports radiating, red tentacle-like appendages.

The anemone stinkhorn first appears as an inconspicuous egg-like structure that could be mistaken for a puffball. But beneath the white outer layer is a strange jellylike goop that contains a compressed ball of tissue called the receptacle, upon which the spore mass, or gleba, develops. Give the "egg" a few days, and its forked red "tentacles" burst forth—often explosively. That's when things start to stink, like old roadkill. The tentacles are soon coated by the malodorous spore mass, a trick to attract flies and other spineless creatures, such as beetles and slugs, that find the slimy slurry sumptuous.

Many fungi have relationships with insects, but the relationships vary in terms of who reaps the rewards. In some pairings, as we'll see in a few hours, parasitic fungi exploit their hosts. In others—such as that of chi-ngulu-ngulu, which we met this morning, and its termite companions—both fungus and insect benefit from the deal. In other equitable fungus-insect unions, little arthropods called springtails seek shelter among a mushroom's lamellae and then pay their dues by leaping around via special taillike levers called furculae, spreading spores as they go. Likewise, slugs graze mushrooms with their raspy, tonguelike radulae and defecate a spore or two here and there. The relationship between the anemone stinkhorn and its visitors also appears to satisfy all parties, except of course unwary humans who dare to venture within sniffing distance.

It's no shock that the anemone stinkhorn

was the first fungus to have an official scientific description in Australia. This flamboyant fungus caught the eye—or perhaps the nose—of French naturalist Jacques Labillardière in 1792. He wasn't seeking the stinkhorn but the lost ships of La Pérouse, which had vanished on an earlier expedition. The Frenchman was "agreeably surprised" with his discovery. Perhaps he had a blocked nose that day.

While humans perceive fungus scents differently, people usually agree that the anemone stinkhorn reeks of rotting flesh. Most of its fellow stinkhorns—some hundred-odd species within the phalloid group—similarly emit strong odors. This suite of stinkhorn scents, all irresistible to the fungi's arthropod assistants, ranges from fecal to floral, with some having a metallic or chemical note. Other fungi release vile odors too, such as the sulfur knight (*Tricholoma sulphureum*), which smells like coal gas. Another, aptly known as the little stinker (*Collybiopsis affixus*), recalls boiling cabbage or a baby that needs changing. The mousepee pinkgill (*Entoloma inca-num*) mimics its obnoxious namesake, and the weeping milkcap (*Lactifluus volemus*) is famously fishy. Many fibrecaps (*Inocybe*) smell like semen, yet when people are asked to describe its odor, they often struggle to pinpoint it.

Originating in Australasia, the anemone stinkhorn has hitchhiked its way to various other destinations, including South Africa, Uruguay, Costa Rica, and the southeast United States. We know little about how it dispersed over such vast distances. But because the species often pops up in garden beds, it's likely that some individuals were transported as propagules hidden in soil—perhaps thanks to the horticultural industry.

When I began teaching fungal ecology almost three decades ago, I'd spend the evening before a workshop setting up a display of various fungi. While out collecting, I came across the anemone stinkhorn in its immature "egg" stage. Excited by my discovery, I placed it in prime position on the

specimen table. As I opened the door the following morning, I was met—to my horror—with
not only the buzzing chorus of what seemed like
Australia's entire blowfly population, immersed
in their favorite fungal fare, but also a vile stench
that permeated the entire room. Collecting
stinkhorns for a fungus display was a mistake
I've only ever made once. While it's an enchanting species, it's one to be enjoyed only for fleeting
moments!

Hairy Nuts Disco

Lanzia echinophila

(EUROPE)

It's early evening in the woods, and the temperature is dropping fast. It's time to build a fire and roast those foraged chestnuts. As we crack open the prickly chestnut husks, we notice something unexpected inside—the tiny cryptic cups of the hairy nuts disco.

These fungi are "spine-loving," as suggested by the specific epithet *echinophila*, growing only on the spiky or hairy husks of chestnuts and sometimes on acorns. What does it mean for a fungus to have such a specific habitat, and how might the limited lodging options affect its survival? What other unexpected habitats have their own personalized fungal inhabitants?

The hairy nuts disco forms miniature cups about the size of a match head or grain of rice, although sometimes reaching the size of a pea. The sporing bodies start out not quite spherical and flatten as they grow. Orange at first, they gradually turn purplish then reddish brown. If you've got good eyes and look beneath the cups, you'll see that they often have tiny stalklike stipes. The hairy nuts disco's generic name, *Lanzia* (*lani-* meaning "wool"), refers to the hairiness of this stem.

Mycologists have recorded the hairy nuts disco across mainland Europe, from the southern Mediterranean to southern Scandinavia. At the time of writing, there are fewer than fifty records for the hairy nuts disco in the digital data repository iNaturalist. Does that mean it's rare? Not necessarily. Small and secretive fungi, especially those (like the hairy nuts disco) that live in restricted habitats, are less likely to be noticed, let alone cataloged, unless observers are in the habit of examining chestnut husks and getting their fingers pricked.

Many fungi can grow in a broad range of habitats, but some, like the hairy nuts disco, are more circumscribed. Both generalist and specialist fungi have enzymes that can dismantle the polymers lignin and cellulose in organic plant material to extract nutrients. Some fungi excel at one or the other, but many can break down both to some extent, allowing them to switch food sources and hence expand the range of habitats in which they're able to live. The loss or fragmentation of their habitats, however, can threaten their existence. Reductions in the amount and diversity of wood—both living and dead, young and old (but especially old)—are a major threat to many fungus species and call for the protection of the remaining old-growth forests.

Many fungus species have a wide "niche breadth"—that is, the diversity of resources at its disposal and the range of environmental conditions it can tolerate. The tenacious honey fungus we met at 2 AM, for example, is highly adaptable and can colonize a variety of plants, especially those that are ill and dying, giving it great geographic and ecological scope. Some fungus species, however, require a specific type of habitat in specific environmental conditions, such as old, unburnt wood. The hairy nuts disco is one such species, choosy when it comes to its already limited habitat: Not satisfied with just any chestnut, it prefers one of a vintage variety, preferably nicely rotted.

Some of the hairy nuts disco's relatives are also picky about where they set up camp. The tiny yellow cup fungus (*L. luteovirescens*) grows on maple petioles, which attach a tree's leaf to its stem. Another species known as the black tack (*L. lanaripes*) because of its resemblance to a dark, flat-headed nail grows on old wet and rotting wood, often with moss, in southern Australia. Cup and disk fungi described as coprophilous, or dung loving, are found in the scat of certain herbivores. Fungi in other groups also have special preferences. The parasitic beech orange (*Cyttaria gunnii*), for example, is specific to myrtle beech trees. The vegetable caterpillar, which you'll meet at 10 PM, targets a particular type of caterpillar to colonize.

When chestnuts are cultivated for human consumption and stored en masse, they're vulnerable to infection by a wide spectrum of microfungi (like *Penicillium*), which can reduce their quality and monetary value. The high nutrient content and moisture levels of chestnuts make them the perfect targets for colonization. While many fungi are under investigation for their crimes against commercially produced chestnuts, the hairy nuts disco appears to be minding its own business and not upsetting anybody.

The hairy nuts disco reminds us that hidden gems can be found in the most unexpected places, as well as the most expected ones—that is, if we know where to look.

7 PM

Red-Green Truffle Milky

Lactarius rubriviridis

(NORTH AMERICA)

Something scurries and snuffles in the under-growth. Quick, where's the flashlight? A stocky, low-slung creature swaggers into view. But it's nothing to worry about; it's just a badger, unin-terested in us and focused on finding one of its favorite foods—truffles.

The famous Périgord truffle, perhaps the most well-known truffle worldwide, is a gour-met fungal delight, especially in France and Italy. Hundreds, maybe thousands, of other truffle species exist, however, and they provide tasty fare for many animals. Animals other than badgers and humans—including chipmunks and squirrels, voles and mice, deer and wild boars, some birds, and countless invertebrates—also feast on truffles. Mammals eat a great range of different fungi. Mycologists who analyzed the feces of the American red squirrel found that it devours an astonishing eighty-nine species of fungi, including many types of truffles. Townsend's chipmunks eat up to eighty-one different species of truffles alone. Some truffles have high concentrations of nutrients and vitamins, and squirrels know to stash this vital food supply, often in unexpected places. One squirrel accumulated a cache of fifty truffles in a deserted robin's nest.

Attracting hungry animals is essential for truffle reproduction. Truffles differ from the other fungi we've encountered so far in that they produce their sporing bodies underground and therefore can't rely on the wind for spore dispersal. How do they advertise themselves as tasty snacks? Truffle fungi release special scents to catch the attention of mammals who will dig them up, gulp them down, then scamper off and defecate them farther away. These funky underground lumps sometimes have pheromonal odors that appeal to mammals whose mates perhaps have a similar scent. Truffles benefit not only from the greater distances over which mammals disperse their spores but also from the nutritious animal poop, which acts as a ready-made starter kit for germination. It's a win-win situation—the mammals score a tasty meal, and the truffle spores get a free ride.

Mycologists think that underground sporing bodies are an evolutionary adaptation for coping with dry conditions. By retreating to the more consistent and predictable conditions of the subterrain, these fungi avoid exposure to the sun and wind. Truffle fungi appear worldwide but are most diverse on dry continents like Australia, where more than forty different mammals, including marsupials such as bandicoots and potoroos, help them disperse their spores.

If you slice the red-green truffle milky open, you'll see some clues to its aboveground past. The milky's inside is full of brain-like convolutions with pinkish-orange pits and holes. The convolutions are crumpled lamellae that are no longer necessary for aboveground spore dispersal. You'll also see what appear to be veins, probably the remnants of a vestigial stipe that is no longer needed to hoist the sporing body above the soil surface and into the wind.

The milky presents us with another conundrum. Most truffles hold their spores inside themselves, relying on animals to disperse them rather than actively discharging them (as we saw at 3 PM). But the milky seems to be keeping its options open; it can also forcibly eject its spores. Could these two features reflect an intermediate stage of the shift between an aboveground and belowground existence?

We know little about this fungus because it's incredibly rare, and researchers have only found it in three places—two in California and one in Oregon. Mycologists think that the milky, like most truffles, is mycorrhizal, growing with pine trees. The biggest threat to its survival is the loss of its pine partners. Despite mycologists' yearly attempts to relocate it, the milky hasn't been spotted since 2001.

The truffles, trees, and creatures represent another compelling example of an interkingdom relationship. Understanding the role of truffles in the diets of small animals highlights the interdependence of organisms and how together they help maintain the vitality of our forests. As the sun sets on our mushroom day, I can't help but wish I could "speak badger" so that these furry friends might tell us where the milky is hiding.

Horn Stalkball

Onygena equina

(WORLDWIDE)

As the moon slips behind clouds and the camp-fire embers burn low, there are squeals and squirming; the ghost stories have begun. Dead bodies feature in these eerie tales, but the true story of a corpse-eating fungus is even more compelling.

Most fungi eat plants. But not sarcophagus fungi. Although the word might conjure images of the pharaohs of ancient Egypt, it has more gruesome origins—with *sárx* meaning "flesh," and *phageîn* "to eat," in ancient Greek. Many sarcophagus fungi feed on the less digestible parts of animal remains. In Australia, for example, the ghoul fungus (*Hebeloma aminophilum*) has a taste for urea and is notorious for its grisly habit of pushing up between animal bones. In the northern hemisphere, the corpse finder (*H. syrjense*) is also true to its name. All in all, about forty fungi fruit more actively in the presence of ammonia, a by-product that is released when organic matter containing nitrogen—like cadavers and animal waste—decomposes.

When an animal dies, it rarely rests in peace. Carrion-craving creatures and other scavengers, such as vultures and hyenas, swiftly move in to feast. Decomposing bodies also invite invertebrates like flies, mites, worms, and beetles, as well as many types of microbes, including bacteria and protozoa. These cadaver-associated organisms constitute the necrobiome. Microbial communities in the necrobiome emerge from the conditions and environment of both the body's past life and its death.

Although creatures quickly clean up a corpse's flesh, some body parts persist, especially bones, teeth, nails, hooves, horns, hair, and feathers. These contain a hard substance—a tough structural protein called keratin. To see how hard keratin is to digest, look no further than the pellets of birds such as owls, hawks, and eagles. More ghastly than a cat's hair ball, these regurgitated lumps contain bones, hair, feathers, and other keratin-containing body parts from the bird's recent meals.

What to do with these leftovers? Enter the horn stalkball—a fungal keratin specialist. Its specific epithet, *equina*, means "horse," and it feeds on herbivores' hooves and horns. Its tiny, whitish to cream-colored sporing body is seldom taller than the width of your little finger. On top of its thick stipe is a flattened ball full of pale, reddish-brown powdery spores. At maturity, the peridium (skin) of the ball breaks, releasing spores to the wind. The sporing bodies may be tiny, but they're numerous: Hundreds adorn old hooves and horns, appearing a little like an old man's stubble. A related keratin-loving species, the feather stalkball (*O. corvina*), dines on damp old feathers and pops up in birds' nests. Another fungus (*Arthroderma curreyi*) not only specializes in deconstructing feathers but fancies tennis balls, so long as they're rotting. Game over.

Most fungi that grow on dead bodies, including human corpses, are microfungi—molds, such as *Penicillium*, *Aspergillus*, and *Mucor*; and yeasts, such as *Candida*. Some fungi are more impatient and grow on bodies while they're still living, causing inflammation and illnesses like ringworm. They inhabit the outer layers of the skin, hair, and nails, reflecting their affinity for keratin. Humans aren't the only animals prone to flesh-loving fungi. The fungal pathogen *Pseudogymnoascus destructans* causes white-nose syndrome, which invades the skin of hibernating bats and causes them to contract subcutaneous infections while asleep in their caves. Likewise, the chytrid fungus *Batrachochytrium dendrobatidis* infects keratinized cells in the skin of frogs, salamanders, and other amphibians, causing chytridiomycosis—a fatal disease devastating their populations worldwide.

I've met various types of mycologists—some who specialize in particular fungi, others who work in agricultural or medicinal fields—but forensic mycologists are among the more curious. They assist in criminal investigations by using the growth rates of fungi that appear on or near bodies to help determine when a victim died. Forensic mycologists sometimes observe fungal succession, and although that process can provide clues about the time of death, it's not a reliable indicator. In cases of mushroom poisoning—accidental or otherwise—these researchers use fungal spores to positively identify a species as the cause of death.

Despite its appearance in true and imagined stories of death, I find the horn stalkball strikingly beautiful. It isn't as macabre as you might think and adds a lovely decorative flourish to an abandoned horse hoof or sheep horn. Other creatures find it appealing too—the fetid fungus, which itself reeks like a cadaver, probably helps attract flies to old remains. And this may actually be one way that the horn stalkball finds corpses in the first place. The red-eyed and bristly-bottomed flesh flies from the family Sarcophagidae spend their time around corpses, sopping up bodily fluids—and possibly horn stalkball spores. Along with the spores, these flies may carry leprosy. They can also cause blood poisoning, so they don't have too many fans outside our ghoulish group of fungi.

This tiny fungus—with its peculiar penchant for body parts—still contributes to life in its own way, tidying up, recycling organic matter, and enriching campfire stories.

Inky Cap

Coprinopsis atramentaria

(WORLDWIDE)

The call comes through to poison control at 9:03 PM. The man on the other end of the line ate mushrooms for dinner and isn't feeling well. He describes to the toxicologist the clump of grayish-white mushrooms that he'd found along the track in the woods behind his home. It was only about fifteen minutes after swallowing the last mouthful of his meal that a growing nausea distracted him from conversation with his wife. He loosened his shirt collar as he broke into a sweat, but his agitation did not subside. A disconcerting feeling of dizziness crept in. It was only when his wife, flushed in the face and short

of breath, complained of a metallic taste in her mouth that he sought help.

The toxicologist takes note of the symptoms and asks if he has a photograph of the mushrooms. Fortunately, he does, and she is quick to identify the culprit. To the caller's surprise, however, it wasn't the mushrooms that had caused the poisoning but the wine.

It turns out that he had foraged the inky cap. Inky caps are edible—and tasty—until alcohol gets involved, as suggested by another common name for the species, the tippler's bane. The man had poured himself a fine glass of merlot. Some foods don't pair well with a red, but the inky cap may pair too well. It contains a mycotoxin called coprine. Coprine heightens the body's sensitivity to ethanol, working similarly to the anti-alcoholism drug disulfiram (also known as Antabuse). It affects how our bodies break down alcohol by inhibiting the function of an important enzyme. So, when someone consumes the inky cap with alcohol, they experience alcohol poisoning. An array of unpleasant symptoms are possible, including nausea, vomiting, sweating, facial reddening, a blotchy rash of the body, headache, malaise, metallic taste of the tongue, difficulty breathing, agitation, vertigo, dizziness, muscle twitching, heart palpitations, and tingling in the limbs. The intensity of the symptoms is relative to the amount of alcohol consumed. In rare circumstances, cardiac arrhythmias—such as atrial fibrillation—can occur, or the consumer can lapse into a coma.

The good news, as the caller discovered, is that symptoms usually subside within a few hours if you put the cork back in the bottle. The less-good news is that you need to keep it there for a few days, as symptoms may occur if you partake within three days of (before or after) eating inky caps. This is probably why the man's wife—who had declined a glass at dinner but had shared a bottle of wine the previous evening—also suffered symptoms of poisoning.

The inky cap grows throughout the northern hemisphere, including in North America, Europe, and Asia. It has also found its way to South Africa, Australia, and New Zealand. It's a ruderal species, growing in human-disturbed habitats, such as woodland tracks, parks and gardens, roadside verges, and vacant lots; it can even push its way through tarmac and lift paving stones. Ruderal species increase their distribution when their spores and propagules are unintentionally dispersed—with a little help from *Homo sapiens*. You'll also find the inky cap in meadows, grasslands, and pretty much anywhere there's buried wood or other organic matter on which this fungus feeds, such as old stumps and dead roots.

The mushrooms can appear individually, but you're more likely to find them in large clusters. The caps start out lead gray or grayish brown, often with a central disk of smooth, flattened scales on top and faint grooves running down their lengths. They are shaped like bullets when young, then expand into bells, often becoming tattered along the edges until "melting"—or deliquescing—into an inky black slop. Other ink caps do this too, as we saw with the lawyer's wig at 10 AM. The specific epithet *atramentaria* comes from the Latin word *atramentum*, which refers to a black or very dark substance, like the inky spore mass.

If you're a teetotaler, you can enjoy these flavorsome mushrooms freely. But you must harvest them and get them in the pan fast, before they dissolve in your basket or, worse, on the back seat of your car—as they did in mine. You could also collect a few to make yourself some ink. Just boil them with a little water and some cloves. When you're done, perhaps write a note reminding others to forgo the whiskey nightcap with their ink cap.

Vegetable Caterpillar

Ophiocordyceps robertsii

(AUSTRALIA AND NEW ZEALAND)

Near the soil surface, there's movement, a tiny disturbance. We hold our breath, sit still, and watch. Under the light of a waning moon, a caterpillar emerges from the subterrain. It's hunting for dinner. The hungry, herbivorous caterpillar munches its way through blades of grass. It seems happy enough, but deep within its digestive tract, a fungus will not only disrupt its feast but spell the end of its life.

Vegetable caterpillars are a group of ento-mopathogenic fungi—that is, fungi that colo-

nize and kill insects. They have evolved a suite of astonishing strategies and tactics, exploiting the spineless bodies of their prey for food and spore dispersal. The name *vegetable caterpillar* is misleading, as these fungi don't just parasitize caterpillars but a great range of insects, other arthropods (such as scorpions and spiders), and even the occasional fungus (such as the deer truffle, *Elaphomyces*).

So how does the vegetable caterpillar fungus find its prey? A caterpillar spends most of the day beneath the soil, in the silklined shaft of its burrow. Under the cover of darkness, it emerges to eat grasses and other organic matter—and may accidentally ingest this fungus's spores. Its innards provide the ideal habitat for the spores to germinate and form a mycelium, which then spreads throughout the caterpillar's body. As the fungus feeds, its powerful digestive enzymes liquefy the luckless creature's internal organs. In the process, the fungus kills the caterpillar and transforms it into a fungal mummy, or sclerotium. Once the fungus has had its fill, it sends its reproductive structure out through the head of the caterpillar and up above the soil surface to release more spores. It's easy to overlook these inconspicuous sporing bodies; as brownish, sometimes branched stems about the height of a pencil but thinner, they can be mistaken for twigs.

Most vegetable caterpillars are picky eaters, singling out a particular species or genus of arthropods for their menu. Many enter the body cavities of their hosts, attracting sci-fi writers and zombie aficionados with this ghoulish behavior. But beyond human fantasies, vegetable caterpillars play a vital role in regulating ecosystems.

When forests are affected by forestry, fires, or climate change, some opportunistic species take advantage of the new conditions to multiply rapidly. A population explosion of one species can deplete resources for others or alter the availability of particular habitats, thereby disrupting forest dynamics. Because most vegetable caterpillars are selective in the invertebrates they target, they can help keep ecosystems stable by preventing any one species from dominating.

The kingdom Fungi is a mysterious realm full of unanswered questions. It's easy to make assumptions and draw false connections in our attempt to justify them. It's not surprising that the bizarre forms of fungal life have prompted some fanciful explanations across the ages. Two centuries ago, this unusual union between a fungus and an invertebrate had scientists stumped, with one proposing that the vegetable caterpillar was an insect that transformed into a plant. Some cultures recognize vegetable caterpillars—with their seemingly inexplicable life histories—as curios or objects of superstition. Chinese, Tibetan, and various other Asian cultures esteem the fungus for its possession of potent chemicals with alleged health benefits.

Ophiocordyceps robertsii also holds historical and cultural significance in New Zealand—and not simply because in 1836 it was the first vegetable caterpillar to be scientifically described in Australasia. Known as *awheto* to the Māori people, this fungus was dried and burned into a charcoal, then pulverized and mixed with bird fat and water to produce a pigment for *tā moko*, or traditional facial tattooing.

Mycologists studying the coevolutionary relationships between parasites and their hosts know that these fungi can manipulate their prey to aid in spore dispersal. Vegetable caterpillars are specialized fungi that have evolved an array of chemicals to alter the physiology and behavior of their hosts. Some vegetable caterpillars produce an ergot alkaloid that causes staggers syndrome in cows, but we do not yet understand how their arthropod prey respond. There's no need for alarm, though. Vegetable caterpillars don't infect vertebrates, which is a relief after the 4 AM show of what ergot alkaloids can do to the human body. It's getting late, so close your eyes and try not to dream that you're a caterpillar.

11PM

Witches Cauldron

Sarcosoma globosum

(EUROPE AND NORTH AMERICA)

In folklore, witches operate under the cover of night, stirring their cauldrons and brewing trouble. By the light of our headlamps, we make out a miniature cauldron too small for any witch to stir. Perhaps the fungus itself is a clairvoyant or seer; indeed, it just might tell us something about the future.

Some fungi are finicky about their habitats. The witches cauldron likes old, mossy nutrient-rich forests next to waterways, not the managed plantation forests that are replacing its former

abode. These plantations haven't yielded any witches cauldrons, and as the practice of clear-cutting continues, only those cauldrons that live in reserves or other protected areas are likely to survive.

Milder winters, wetter springs, and changes in seasonal flooding patterns—all trends associated with climate change—are affecting witches cauldrons around the globe. Sweden is still a major stronghold for the species, and its numbers in Finland appear to have increased over the last three decades. But its once abundant Norwegian contingent is in steady decline, and its German, Slovak, and Lithuanian populations are already thought to be extinct. In other places, the fungus's appearance is unpredictable. It used to be rare in Estonia; now, it's relatively common—but only in some years.

Interpreting the distribution of the witches cauldron is tricky. What might appear to be an increase in population size, such as we see in Finland, may simply be an increase in observation activity. Conversely, foragers tend to look for fungi in fall; as a spring species, the witches cauldron may not be rare but overlooked in favor of orchids and birds. These trends disguise any real changes occurring in natural populations.

Even when climate change doesn't drive extinction, temperature increases affect the phenology of fungi. The witches cauldron, like the midnight disco, typically fruits in the spring. Its "cauldrons" develop beneath the snow, revealing themselves only as it melts away and then lingering for a couple months. They can swell to the size of your palm, flattening out over time and often becoming wrinkly at maturity. The inner surface of the cup is coated with a transparent gel, while the blackish-brown exterior starts out velvety but toughens with age.

Over the last fifty years, mycologists in the United Kingdom have recorded changes in the seasonal patterns of fungus fruiting. In many parts of Europe, the "mushroom season" — typically from late summer through early winter—has been starting earlier and lasting longer. Some fungus species that previously fruited only in fall now also fruit in spring. The interplay of local climates and conditions and the specialized requirements of different fungi, however, remains poorly understood, and mycologists are trying to determine how shifts in fruiting aboveground relate to fungal activities below the soil surface.

To unravel these mysteries, we need to spend more than a day with fungi. There are too few mycologists to face the enormous task of understanding and protecting this kingdom. Perhaps you might join us. In recent years, "citizen scientists" have helped monitor fungi, contributing records to online data repositories and increasing our awareness of species distribution. The witches cauldron is a perfect candidate for citizen science projects because it is distinctive, with few look-alike species, and relatively easy to identify.

It seems that the witches cauldron is sensitive to its environmental conditions, which makes it a useful indicator of change. And because it is long-lived—its mycelium might exist for several decades—changes in its abundance and distribution reflect environmental changes over longer periods of time. So, as we end our mushroom day, we might reflect on how to protect these prophetic fungi for tomorrow, and always.

Epilogue

It was difficult to select just twenty-four fungi for us to meet on our mushroom day. With fungi occupying such a diverse kingdom of life, there's a very large group of candidates to choose from. Next time we might need an entire mushroom year!

Far fewer fungi are known and named relative to animals and plants, and so much about them awaits discovery. As we learn more about fungus biology and distribution, and how they respond to light, many more interesting contestants will likely reveal themselves. I'm working on my short list of favorites and have narrowed it down to just a few hundred species.

The different voices we heard throughout our mushroom day—mycologists, ecologists, foragers, conservationists, surveyors, toxicologists,

First Nations peoples, mushroom inspectors, and myriad other mycophiles—all contribute to a broader understanding and appreciation of fungi. But there are other voices too, especially young ones, whose imaginations are unencumbered by the presumptions of adult minds and who endlessly surprise me.

Early one evening a few years back, I was giving a talk on fungi in a tiny southeastern Australian town. I was describing the perils of the folkloric tales about edible and toxic mushrooms that have emerged across the globe and over the centuries. Perhaps you've heard some of these myths too, like how a poisonous mushroom blackens a silver spoon or turns garlic blue. Or how a mushroom is edible if it has pink lamellae or you can peel its cap. Many of these adages arose during a time when fungi were not well understood, and few hold truths—especially when transported from one place to another. It would certainly make foraging easier if there were such reliable guides, but that's not the case.

Anyway, I asked the audience whether feeding a mushroom to another mammal might be a good way to test edibility for humans, given that we too are mammals.

"What say we feed the mushroom to a dog, then carefully observe the animal for a couple days to ensure that it shows no signs of poisoning? Surely, if the dog is OK, we can eat that mushroom," I proffered.

Some audience members contemplated my proposition, while others vehemently shook their heads and cried out, "No!"

"But why not?" I taunted.

Then, a small girl about four years old stood up and stepped forward. The audience quieted. She glared at me and, with an air of frustrated disbelief, bellowed, "No!"

"No? But why not?"

She paused then, narrowing her eyes, retorted, "We can't eat the mushroom because it's in the dog."

It wasn't the response I was expecting. The

audience roared with laughter. What can I say other than that she was right? I certainly wasn't about to suggest that we turn the dog upside down and shake it until the mushroom falls out. Next time I'll think more carefully about how I phrase my questions.

With every foray into the forest and every talk, I discover more about mushrooms—and more about humans. All these experiences and encounters offer new insights into and perspectives on fungi. I secretly hoped that the young girl's annoyance with the limitations of my adult thinking might inspire her to pursue a career in mycology. I'll be keeping an eye out for her. And I promise never to feed potentially poisonous mushrooms to our canine friends.

Fungi are fascinating and a little bit addictive. So, I should probably warn you that once your mind is fungally infected, there's no going back. Thank you for joining me today, and I hope that every day can be a mushroom day.

Acknowledgments

I extend my sincere thanks to Joseph Calamia from the University of Chicago Press for the invitation to write this book and for his astute guidance. Joe has been an absolute joy and great fun to work with in every way. Thank you to the University of Chicago Press team, especially Lily Sadowsky for her generous and judicious editing and designer Jill Shimabukuro. Working with the delightful Stuart Patience—and watching his wonderful and whimsical fungus illustrations materialize on the page—was as amazing as the emergence of mushrooms themselves. Special thanks to Valérie Chételat for her clever and quirky contributions at the conceptual stage and for commenting on drafts.

Thank you to mycologists Scott Redhead, Steve Trudell, Britt Bunyard, Lorelei Norvell, Anders Dahlberg, Greg Mueller, and Annu Ruotsalainen and lichenologist R. Troy McMullin for their invaluable advice and generous assistance in ensuring mycological and technical accuracy. Also, enormous thanks to the two anonymous scientific reviewers who greatly improved the manuscript.

I'm deeply appreciative of the invitation from Christof Mauch, director of the Rachel Carson Center for Environment and Society at Ludwig-Maximilians-Universität in Munich, to partake in a Volkswagen Foundation Visiting Professorship in the second half of 2024. This provided me with some unexpected time and funds—and loads of inspiration—to bring the manuscript to completion.

I'm grateful to Simone Sylvestre, Sergio Ferrero, Barbara Thüler, Anders Hirell, Sarah Riley, and Josh Wayborn for sharing their fungus stories and experiences and allowing me to include them in this book.

And thank you to my many adventurous field companions, with whom I've experienced the wonders of forests and fungi around the world.

Further Reading

Boddy, Lynne, and Ali Ashby. 2023. *Fungi: Discover the Science and Secrets Behind the World of Mushrooms*. Dorling Kindersley.

Bunyard, Britt A. 2022. *The Lives of Fungi: A Natural History of Our Planet's Decomposers*. Princeton University Press.

Kendrick, Bryce. 2017. *The Fifth Kingdom: An Introduction to Mycology*. 4th ed. Hackett Publishing.

Lücking, Robert, and Toby Spribille. 2024. *The Lives of Lichens: A Natural History*. Princeton University Press.

Maser, Chris, Andrew W. Claridge, and James M. Trappe. 2008. *Trees, Truffles, and Beasts: How Forests Function*. Rutgers University Press.

Moore, David, Geoffrey Robson, and Anthony Trinci. 2000. *21st Century Guidebook to Fungi*. Cambridge University Press.

Petersen, Jens Henrik. 2013. *The Kingdom of Fungi*. Princeton University Press.

Pouliot, Alison. 2023. *Meetings with Remarkable Mushrooms*. University of Chicago Press.

Sheldrake, Merlin. 2020. *Entangled Life*. Bodley Head.

PREFACE

Fischer, Reinhard, Jesus Aguirre, Alfredo Herrera-Estrella, and Luis M. Corrochano. 2016. "The Complexity of Fungal Vision." *Microbiology Spectrum* 4 (6): 1–22. https://doi.org/10.1128/microbiolspec.funk-0020-2016.

Herrera-Estrella, Alfredo, and Benjamin A. Horwitz. 2007. "Looking Through the Eyes of Fungi: Molecular Genetics of Photoreception." *Molecular Microbiology* 64 (1): 5–15.

Yu, Zhenzhong, and Reinhard Fischer. 2019 "Light Sensing and Responses in Fungi." *Nature Reviews Microbiology* 17 (January): 25–36.

MIDNIGHT: MIDNIGHT DISCO

Diez, Jeffrey, Håvard Kauserud, Carrie Andrew, et al. 2020. "Altitudinal Upwards Shifts in Fungal Fruiting in the Alps." *Proceedings of the Royal Society B* 287 (1919): 20192348. https://doi.org/10.1098/rspb.2019.2348.

Vitasse, Yann, Sylvain Ursenbacher, Geoffrey Klein, et al. 2021. "Phenological and Elevational Shifts of Plants, Animals and Fungi Under Climate Change in the European Alps." *Biological Reviews* 96 (5): 1816–35. https://doi.org/10.1111/brv.12727.

1AM: GHOST FUNGUS

Herring, Peter J. 1994. "Luminous Fungi." *Mycologist* 8 (4): 181–
  83. https://doi.org/10.1016/S0269-915X(09)80193-6.

Weinstein, Philip, Steven Delean, Tom Wood, and Andrew Aus-
  tin. 2016. "Bioluminescence in the Ghost Fungus *Omphalotus nidiformis* Does Not Attract Potential Spore Dispersing
  Insects." *IMA Fungus* 7 (2): 229–34. https://doi.org/10.5598/
  imafungus.2016.07.02.01.

2AM: HONEY FUNGUS

Mihail, Jeanne, Landon Bilyeu, and Sara R. Lalk. 2018. "Bioluminescence Expression During the Transition from Mycelium to Mushroom in Three North American *Armillaria*
  and *Desarmillaria* Species." *Fungal Biology* 122 (11): 1064–68.
  https://doi.org/10.1016/j.funbio.2018.08.007.

3AM: DEVIL'S BOLETE

Gachet, Christian, Rachid Ennamany, Olivier Kretz, et al.
  1996. "Bolesatine Induces Agglutination of Rat Platelets
  and Human Erythrocytes and Platelets In Vitro." *Human
  and Experimental Toxicology* 15 (1): 26–29. https://doi.
  org/10.1177/096032719601500105.

Landi, Nicola, Hafiza Z. F. Hussain, Paolo V. Pedone, Sara Ragu-
  cci, and Antimo Di Maro. 2022. "Ribotoxic Proteins, Known

as Inhibitors of Protein Synthesis, from Mushrooms and Other Fungi According to Endo's Fragment Detection." *Toxins* 14 (6): 403. https://doi.org/10.3390/toxins14060403.

4AM: ERGOT

Florea, Simona, Daniel G. Panaccione, and Christopher L. Schardl. 2017. "Ergot Alkaloids of the Family Clavicipitaceae." *Phytopathology* 107 (5): 504–18. https://doi.org/10.1094/PHYTO-12-16-0435-RVW.

Smakosz, Aleksander, Wiktoria Kurzyna, Michał Rudko, and Mateusz Dąsal. 2021. "The Usage of Ergot (*Claviceps purpurea* [fr.] Tul.) in Obstetrics and Gynecology: A Historical Perspective." *Toxins* 13 (7): 492. https://doi.org/10.3390/toxins13070492.

5AM: HOU TOU GU

Wang, Mingxing, Yang Gao, Duoduo Xu, Tetsuya Konishi, and Qipin Gao. 2014. "*Hericium erinaceus* (Yamabushitake): A Unique Resource for Developing Functional Foods and Medicines." *Food and Function* 5 (12): 3055–64. https://doi.org/10.1039/c4fo00511b.

Abesha, Emnet, Gustavo Caetano-Anollés, and Klaus Høiland. 2003. "Population Genetics and Spatial Structure of the Fairy Ring Fungus *Marasmius oreades* in a Norwegian Sand Dune Ecosystem." *Mycologia* 95 (6): 1021–31.

Hiltunen, Markus, Sandra Lorena Ament-Velásquez, Martin Ryberg, and Hanna Johannesson. 2022. "Stage-Specific Transposon Activity in the Life Cycle of the Fairy-Ring Mushroom *Marasmius oreades*." *Proceedings of the National Academy of Sciences* 119 (46): e2208575119. https://doi.org/10.1073/pnas.2208575119.

Boa, Eric. 2004. *Wild Edible Fungi: A Global Overview of Their Use and Importance to People*. Food and Agriculture Organization of the United Nations. http://www.fao.org/3/y5489e/y5489e00.htm.

Pouliot, Alison, and Tom May. 2021. *Wild Mushrooming: A Guide for Foragers*. CSIRO Publishing.

Salerni, Elena, and Claudia Perini. 2004. "Experimental Study for Increasing Productivity of *Boletus edulis* s.l. in Italy." *Forest Ecology and Management* 201 (2–3): 161–70. https://doi.org/10.1016/j.foreco.2004.06.027.

8AM: CHI-NGULU-NGULU

Aanen, Duur, Paul Eggleton, Corinne Rouland-Lefèvre, Tobias Frøslev, Søren Rosendahl, and Jacobus J. Boomsma. 2002. "The Evolution of Fungus-Growing Termites and Their Mutualistic Fungal Symbionts." *Proceedings of the National Academy of Sciences* 99 (23): 14887–92.

Palma, Jb, and Julean Federizo. 2022. "Phylogenetic Relationships and Biogeographic Distribution of Termitomyces (Lyophyllaceae, Basidiomycota)." *Ecosystems and Development Journal* 12 (2): 85–96.

9AM: VEILED POLYPORE

Kadowaki, Kohmei. 2010. "Species Coexistence Patterns in a Mycophagous Insect Community Inhabiting the Wood-Decaying Bracket Fungus *Cryptoporus volvatus* (Polyporaceae: Basidiomycota)." *European Journal of Entomology* 107 (1): 89–99.

Park, Myung Soo, Jonathan J. Fong, Hyun Lee, et al. 2014. "Determination of Coleopteran Insects Associated with Spore Dispersal of *Cryptoporus volvatus* (Polyporaceae: Basidiomycota) in Korea." *Journal of Asia-Pacific Entomology* 17 (4): 647–51. https://doi.org/10.1016/j.aspen.2014.06.005.

Watson, David, and David Shaw. 2018. "Veiled Polypore (*Cryptoporus volvatus*) as a Foraging Substrate for the White-Headed Woodpecker (*Picoides albolarvatus*)." *Northwestern Naturalist* 99:58–63.

10AM: LAWYER'S WIG

Luo, Hong, Minghe Mo, Xiaowei Huang, Xuan Li, and Keqin Zhang. 2004. "*Coprinus comatus*: A Basidiomycete Fungus Forms Novel Spiny Structures and Infects Nematode." *Mycologia* 96 (6): 1218–24.

Luo, Hong, Yajun Liu, Lin Fang, Xuan Li, Ninghua Tang, and Keqin Zhang. 2007. "*Coprinus comatus* Damages Nematode Cuticles Mechanically with Spiny Balls and Produces Potent Toxins to Immobilize Nematodes." *Applied and Environmental Microbiology* 73 (12): 3916–23. https://doi.org/10.1128/AEM.02770-06.

11AM: GOLDEN CHANTERELLE

Egli, Simon, Martina Peter, Christoph Buser, Werner Stahel, and François Ayer. 2006. "Mushroom Picking Does Not Impair Future Harvests: Results of a Long-Term Study in Switzerland." *Biological Conservation* 129:271–76.

Norvell, Lorelei. 1995. "Loving the Chanterelle to Death? The Ten-Year Oregon Chanterelle Project." *McIlvainea* 12:6–25.

Peintner, Ursula, Stefanie Schwarz, Armin Mešić, Pierre-Arthur Moreau, Gabriel Moreno, and Philippe Saviuc. 2013. "Mycophilic or Mycophobic? Legislation and Guidelines on Wild Mushroom Commerce Reveal Different Consumption Behaviour in European Countries." *PLOS ONE* 8 (5): e63926. https://doi.org/10.1371/journal.pone.0063926.

NOON: ARCTIC ORANGEBUSH LICHEN

McMullin, R. Troy. 2023. *Lichens: The Macrolichens of Ontario and the Great Lakes Region of the United States*. Firefly Books.

McMullin, R. Troy, and Daniel Kraus. 2021. "Canada's Endemic Lichens and Allied Fungi." *Evansia* 38 (4): 159–73. https://doi.org/10.1639/0747-9859-38.4.159.

1PM: PENICILLIUM

Crequer, Ewen, Jeanne Ropars, Jean-Luc Jany, et al. 2023. "A New Cheese Population in *Penicillium roqueforti* and Adaptation of the Five Populations to Their Ecological Niche." *Evolutionary Applications* 16 (8): 1438–57. https://doi.org/10.1111/eva.13578.

Gillot, Guillaume, Jean-Luc Jany, Elisabeth Poirier, et al. 2017. "Functional Diversity Within the *Penicillium roqueforti* Species." *International Journal of Food Microbiology* 241:141–50. https://doi.org/10.1016/j.ijfoodmicro.2016.10.001.

Gillot, Guillaume, Jean-Luc Jany, Monika Coton, et al. 2015. "Insights into *Penicillium roqueforti* Morphological and Genetic Diversity." *PLOS ONE* 10 (6): e0129849. https://doi.org/10.1371/journal.pone.0129849.

2PM: BRANCHED SHANKLET

Chachuła, Piotr, and Piotr Mleczko. 2020. "*Dendrocollybia racemosa* (Pers.) R. H. Petersen & Redhead (Tricholomataceae, Agaricales): New Localities and a New Host." *Monografie* 16:45–54.

Lennox, Joanne W. 1979. "Collybioid Genera in the Pacific Northwest." *Mycotaxon* 9:117–231.

Machnicki, Noelle, Leesa Wright, Alissa Allen, et al. 2006. "*Russula crassotunicata* Identified as Host for *Dendrocollybia racemosa*." *Pacific Northwest Fungi* 1 (9): 1–7. https://doi.org/10.2509/pnwf.2006.001.009.

3PM: SANDY STILTBALL

Gargano, Maria L., Giuseppe Venturella, and Valeria Ferraro. 2021. "Is *Battarrea phalloides* Really an Endangered Species?" *Plant Biosystems: An International Journal Dealing with all Aspects of Plant Biology* 155 (4): 759–62. https://doi.org/10.1080/11263504.2020.1779847.

4PM: ANISEED FUNNEL

Boniface, Tony. 2020. "The Use of Odours in the Identification of Mushrooms and Toadstools." *Field Mycology* 21 (1): 28–30. https://doi.org/10.1016/j.fldmyc.2020.01.010.

Rapior, Sylvie, Sophie Breheret, Thierry Talou, Yves Pélissier, and Jean-Marie Bessière. 2002. "The Anise-Like Odor of *Clitocybe odora, Lentinellus cochleatus* and *Agaricus essettei.*" *Mycologia* 94 (3): 373–76. https://doi.org/10.1080/15572536.2 003.11833201.

5PM: ANEMONE STINKHORN

May, Tom, Josephine Milne, Susan Shingles, and Rhys Jones. 2003. *Fungi of Australia.* Vol. 2B. CSIRO Publishing.

Pouliot, Alison. 2018. *The Allure of Fungi.* CSIRO Publishing.

6PM: HAIRY NUTS DISCO

Rodrigues, Paula, Jihen Oueslati Driss, José Gomes-Laranjo, and Ana Sampaio. 2022. "Impact of Cultivar, Processing and Storage on the Mycobiota of European Chestnut Fruits." *Agriculture* 12 (11): 1930. https://doi.org/10.3390/agriculture12111930.

7PM: RED-GREEN TRUFFLE MILKY

Desjardin, Dennis E. 2003. "A Unique Ballistosporic Hypogeous

Sequestrate *Lactarius* from California." *Mycologia* 95:148–55. https://doi.org/10.1080/15572536.2004.11833144.

Sultaire, Sean M., Gian Maria Niccolò Benucci, Reid Longley, et al. 2023. "Using High-Throughput Sequencing to Investigate Summer Truffle Consumption by Chipmunks in Relation to Retention Forestry." *Forest Ecology and Management* 549:121460. https://doi.org/10.1016/j.foreco.2023.121460.

8PM: HORN STALKBALL

Bunyard, Britt A. 2022. *The Lives of Fungi: A Natural History of Our Planet's Decomposers.* Princeton University Press.

Carter, David, and Mark Tibbett. 2003. "Taphonomic Mycota: Fungi with Forensic Potential." *Journal of Forensic Sciences* 48 (1): 168–71.

Hawksworth, David L., and Patricia Wiltshire. 2011 "Forensic Mycology: The Use of Fungi in Criminal Investigations." *Forensic Science International* 206 (1–3): 1–11. https://doi.org/10.1016/j.forsciint.2010.06.012.

9PM: INKY CAP

Heleno, Sandrina A., Isabel C. F. R. Ferreira, Ricardo C. Calhelha, Ana P. Esteves, Anabela Martins, and Maria João R. P. Queiroz. 2014. "Cytotoxicity of *Coprinopsis Atramentaria* Extract, Organic Acids and Their Synthesized Methylated and Glucuronate Derivatives." *Food Research International* 55:170–75. https://doi.org/10.1016/j.foodres.2013.11.012.

10PM: VEGETABLE CATERPILLAR

Beckerson, William C., Courtney Krider, Umar A. Mohammad, and Charissa de Bekker. 2023. "28 Minutes Later: Investigating the Role of Aflatrem-Like Compounds in *Ophiocordyceps* Parasite Manipulation of Zombie Ants." *Animal Behaviour* 203:225–40. https://doi.org/10.1016/j.anbehav.2023.06.011.

Xu, Melvin, Nathan A. Ashley, Niloofar Vaghefi, Ian Wilkinson, and Alexander Idnurm. 2023. "Isolation of Strains and Their Genome Sequencing to Analyze the Mating System of *Ophiocordyceps robertsii*." *PLOS ONE* 18 (5): e0284978. https://doi.org/10.1371/journal.pone.0284978.

11PM: WITCHES CAULDRON

Boddy, Lynne, Ulf Büntgen, Simon Egli, et al. 2014. "Climate Variation Effects on Fungal Fruiting." *Fungal Ecology* 10:20–33.

Ohenoja, Esteri, Maarit Kaukonen, and Anna L. Ruotsalainen. 2013. "*Sarcosoma globosum*—an Indicator of Climate Change?" *Acta Mycologica* 48 (1): 81–88. https://doi.org/10.5586/am.2013.010.

Ruotsalainen, Anna L., Tapio Kekki, Esteri Ohenoja, and Tea von Bonsdorff. 2023. "Increase in *Sarcosoma globosum* Observations Reveals New Fungal Observation Culture." *Fungal Ecology*, 65: 101282. https://doi.org/10.1016/j.funeco.2023.101282.

Index

omonal, 131; sensitivity to, 112–13; smell-alike species, 114; and subjectivity, 113; subsiding with age, 111; sweet, 79, 112; of truffle fungi, 131; vocabulary for, limited in English, 113

Omphalotus nidiformis. See ghost fungus

Onygena corvina. See feather stalkball

Onygena equina. See horn stalkball

Ophiocordyceps robertsii. See vegetable caterpillar

oysters (fungi), 74

P. roquefort. See penicillium

Pachyella violaceonigra. See midnight disco

p-Anisaldehyde, 112

Paracelsus, 34

parasites, 11, 17, 26–27, 31, 74, 101, 118, 127; coevolutionary relationships with hosts, 152; and insects, 147, 149–50. *See also* symbiosis

pathogens, 81–82, 86, 138, 147, 149

pedogenesis, 66

penicillium, 90–96, 90, 127, 138, 172–73; and cheeses, 91–96

Périgord truffles, 129–30

phalloids, 117, 119

pharmaceuticals, 35. *See also* medicines

phenology, 3–5, 156

phosphorescence, marine, 12

photosynthesis, 66, 85–86

Picoides albolarvatus. See white-headed woodpeckers

Pleurotaceae. *See* oysters (fungi)

poison pies, 49

poisonous fungi, 25–26, 33–34, 49–50, 54–55, 87, 108, 114, 139–44, 160, 162. *See also* edibility; ergotism; toxicology; toxins

Poland, 107

pollinators, 4

polypores. *See* veiled polypore

porcino, 23, 51–56, 52, 80, 169; flavor, as unsurpassed, 53; hunted in Italy, xviii

Portugal, 103, 106, 108

predators, 4, 18, 73, 86

propagules, 120, 145

protozoa, 136

Pseudogymnoascus destructans (fungal pathogen), 138

pseudorhiza, 62

pseudoscience, and marketing, 41

puffballs: and anemone stink-horn, 117; desert-stalked, 103; and fairy rings, 47–48; giant,

22, 32, 46, 60, 62, 71–72, 103,
105–6, 112, 117–18, 120, 137, 145,
149, 152; by insects, 67–68; by
invertebrates, 99–100, 130–32,
150–51. *See also* germination;
reproduction
spring season, 3–4, 34, 155–57
squirrels, American red, 130
St. Anthony's fire, 33
stalkballs. *See* feather stalkball;
horn stalkball
Steinpilz (stone mushroom). *See*
porcino
stone mushroom. *See* porcino
Stropharia aeruginosa. See verdigris agaric
subterrain fungi, xiv, xvi, 22,
131, 147
sulphur knight, 119
summer season, 3, 88–89, 100, 103,
156–57
surveyors, 159–60
Sweden, 39, 77–81, 155
Switzerland, 34, 55, 81
Sylvestre, Simone, 23, 28–29
symbiosis: and cooperation,
69, 89; and ecosystems, 69;
interkingdom, 89, 133. *See also*
mutualism; parasites

Taiwan, 40. *See also* China
tattoos, 152
taxonomy, v, 79
technology, and commercial
mushroom production, 40
termites, 57, 59–62, 118
Termitomyces titanicus. See chi-
ngulu-ngulu
Thüler, Barbara, 55
Tibet, 151
tippler's bane. *See* inky cap
tooth fungus, 37
toxicology, v, 34, 141, 143, 159–60.
See also poisonous fungi;
toxins
toxins, v, 25–26, 28, 31, 34–35, 49,
51, 53–54, 56, 73–74, 79, 83, 87,
114, 160; myco-, 143. *See also*
ergotism; poisonous fungi;
toxicology
Tricholoma sulphureum. See sulphur knight
trooping funnel, 48
tropical fungi, 18, 107
truffles, 129–33; hunted, 113; Périgord, 129–30; reproduction, 130;
white, 51. *See also* deer truffle;
red-green truffle milky

EARTH DAY